Microbial Desalination Cells for Low Energy Drinking Water

Microbial Desalination Cells for Low Energy Drinking Water

Edited by

Sergio G. Salinas-Rodríguez,
Juan Arévalo,
Juan Manuel Ortiz,
Eduard Borràs-Camps,
Victor Monsalvo-Garcia,
Maria D. Kennedy and
Abraham Esteve-Núñez

IWA PUBLISHING

Published by **IWA Publishing**
Unit 104 – 105, Export Building
1 Clove Crescent
London E14 2BA, UK
Telephone: +44 (0)20 7654 5500
Fax: +44 (0)20 7654 5555
Email: publications@iwap.co.uk
Web: www.iwapublishing.com

First published 2021

British Library Cataloguing in Publication Data
A CIP catalogue record for this book is available from the British Library

ISBN: 9781789062113 (paperback)
ISBN: 9781789062120 (eBook)

Sergio G. Salinas-Rodríguez, Juan Arévalo, Juan Manuel Ortiz, Eduard Borràs-Camps, Victor Monsalvo-Garcia, Maria D. Kennedy, Abraham Esteve-Núñez (Eds.) (2021). *Microbial Desalination Cells for Low Energy Drinking Water*, London: IWA Publishing. Doi: 10.2166/9781789062120

Contents

List of Contributors

		Chapter authors
Aqualia, Spain	Juan Arévalo-Vilches	1, 4, 5, 8
	N. Hernández-Ibáñez	4, 5
	Vicente F. Mena	4, 5
	Victor Monsalvo-Garcia	1, 4, 5
	Frank Rogalla	2, 4, 5
	Patricia Zamora	2, 4
Fujifilm, Netherlands	Abdulsalam Alhadidi	3
	Maarten Meijlink	3
IHE Delft Institute for Water Education	Maria D. Kennedy	1, 7
	Sonam Jamtsho	7
	Angeles Mendoza-Sammet	1, 7
	Sergio G. Salinas-Rodríguez	1, 7, 8
IMDEA Water, Spain	Abraham Esteve-Núñez	2, 8
	Juan M. Ortiz	1, 2, 3, 8
	Marina Ramírez-Moreno	2, 3
	Pau Rodenas	2, 3

Leitat Technological Center, Spain	Martí Aliaguilla	2, 3
	Eduard Borràs Camps	1, 2, 3, 8
	Pau Bosch-Jimenez	2, 3
	Nathan Bossa	7
	Camilla Delpivo	7
	Maria Diez	7
	M. Faccini	3
	Verónica González Andrés	7
	Laura Huidobro	3
	M. Juan-y-Seva	3
	Sandra Martínez-Crespiera	3
	Sonia Matencio	3
	Daniele Molognoni	3
Oncontrol Technologies, Portugal	Tiago Matias	6
	Pedro Sousa	6
	Francisco Alexandre A. Souza	6
SGL Carbon, Germany	Ruediger Schweiss	3
	Almut Schwenke	3
SimTech Simulation Technology, Austria	Stefan Bergmann	6
	Fatima Dargam	6
	Erhard Perz	6
	Ekaterina Rodionova	6

About the Editors

Dr. Sergio G. Salinas- Rodríguez is Associate Professor of Water Supply Engineering at IHE Delft Institute for Water Education. He is a desalination and water treatment technology professional with experience in Latin America, the Middle East, and Europe. Sergio has a PhD in Desalination and Water Treatment from the Technical University of Delft (Netherlands), an MSc in Water Supply Engineering from UNESCO-IHE Institute for Water Education (Netherlands), a Master's in Irrigation and Drainage and a BSc in Civil Engineering from San Simon Major University (Bolivia). He also obtained the University Teaching Qualification, a qualification of pedagogical competences of university teachers. He has over 50 publications in international peer-reviewed journals, book chapters in top publishing houses, and conference proceedings in the areas of seawater and brackish water desalination, water treatment, water reuse, and natural organic matter characterization.

Sergio is involved in teaching and curriculum development of the MSc Programme in Urban Water and Sanitation (UWS) at IHE Delft. Furthermore he has worked in capacity building and in research and innovation projects, and has been project leader in several of them. He has mentored more than 40 MSc students, co-promoted 3 PhD students, and currently supervises 3 PhD students.

Dr. Juan Arévalo is Project Manager at the Innovation and Technology Department of FCC Aqualia. He has a Degree in Environmental Sciences and obtained his PhD in the Department of Civil Engineering in the University of Granada, in Spain. He was visiting researcher at the University of Montpellier. He worked in the University of Granada until 2012 developing research involving MBR membrane technology optimization, desalination pre-treatments and reverse osmosis and UF membrane autopsies. From 2013 to 2016 he worked in the private company Abengoa Water, working on projects related to membrane technology use as pre-treatment in desalination plants, tertiary treatments in wastewater reuse or industrial water treatment, and leading the membrane laboratory. Since October 2016 he has been working at FCC Aqualia, in the R&D Department developing projects related to new desalination methods. He has worked on 16 research projects in public and private organizations with around 17 publications in international peer-reviewed journals.

Dr. Juan M. Ortiz has a PhD in Electrochemistry: Science and Technology from the University of Alicante (2009). Thesis: 'Water Desalination by Electrodialysis Systems Powered directly by Photovoltaic Solar Energy: Technical Feasibility and Modeling Study'. He is researcher at IMDEA Agua (Madrid, Spain). After 10 years of research at the University of Alicante, in the Applied Electrochemistry and Electrocatalysis Research Group (University Institute of Electrochemistry), he joined IMDEA Water Institute in 2015. His research is focused on the development of microbial electrochemical processes for the treatment of urban and industrial wastewater, and sustainable desalination systems. He has participated in 11 national and 4 European projects, including MIDES (Horizon 2020) WP Leader. He is the author of 20 scientific publications in the area of water technology and four national and international patents.

Dr. Eduard Borràs Camps is a Senior Researcher in the Circular Economy Department at Leitat Technological Center. He holds a Degree in Environmental Sciences (2005) and a PhD in Environmental Science and Technology (2011) from the Autonomous University of Barcelona. His research experience has focused on environmental biotechnology in projects related to bioremediation, waste valorization and microbial electrochemical technologies. He was a visiting researcher at the Laboratory of Environmental Biotechnology in the Institute of Microbiology of the Academy of Sciences of the Czech Republic. He has been Assistant Professor in the Chemical, Biological and Environmental Engineering Department of the Autonomous University of Barcelona. His recent professional career has been focused on the execution and management of microbial electrochemical technologies projects for energy harvesting, nutrient recovery, pollution abatement and drinking water production in 15 international and national R&D projects. Also, he has participated in several conferences and published over 14 peer-reviewed papers.

Dr. Victor Monsalvo is Head of the Eco-efficiency Area in FCC Aqualia. He was appointed in 2019 as Chair of R&D Committee at the International Desalination Association. He has a PhD in Chemical Engineering from University Autonoma de Madrid, and held the position of Assistant Professor in the Chemical Eng. Department from 2005 to 2014, and visiting researcher in different Universities: Leeds (2007), Cranfield (2009), Sydney (2011) and Aachen (2013). He was then appointed as Head of Technological Area – Water in Abengoa Research until 2016. Victor leads an active research team working in water treatment and reuse, biowastes and biogas management, membrane technologies and processes involved in desalination plants, and has been involved in more than 25 research projects (publicly and privately funded). He has (co) authored 6 patents (national and international), 1 utility model, around 60 journal and referred conference papers, 3 book chapters and edited 4 books. He has participated in evaluation committees in the PARTNERSHIPS Programme Joint Applied Research Projects (2011–2014) Ministry of Education, Research, Youth and Sport (Romania) and in the National Committee for Evaluation of 'Proyectos de investigacion fundamental no orientada' 2012 for the Ministry of Economy and Competitiveness (Spain).

Dr. Maria D. Kennedy is Professor of Water Treatment Technology at IHE Delft. She has over 28 years of experience in education, research, consulting and capacity development in water treatment. During the last 28 years, she has been involved in the supervision of over 200 MSc participants and 22 PhD research fellows in the areas of water quality, groundwater treatment, disinfection, advanced oxidation, surface water treatment, desalination and membrane related technology, natural treatment systems, water reuse, water transport and distribution and biological stability. She has over 150 publications in peer-reviewed journals (h index > 34), and she has edited several books/book chapters on various aspects of water treatment.

Professor Maria Kennedy was also involved in several large EU/Horizon 2020 research projects such as EU MEDINA, EU TECHNEAU, EUROMBRA, H2020 MIDES and EU India H20.

Professor Maria Kennedy is a past president of the European Desalination Society (EDS), and Chairman of the board of directors. She was also a member of the Science and Technology Board of the EU Joint Programming Initiative (JPI), and she is a jury member of several prestigious international technology events such as the USAID Water for Food Desalination Prize (2014–2015), the Oman Humanitarian Desalination Challenge and the Aquatech Innovation Award.

Dr. Abraham Esteve-Núñez has a PhD in Biochemistry from the University of Granada (2000) with an extraordinary award, and he is a tenured Professor at the University of Alcalá (UAH) and associate researcher at IMDEA Agua (Madrid, Spain), where he leads the Bioe Group (www.bioelectrogenesis.es). After postdoctoral stays at the University of Massachusetts-Amherst (USA), and at the Center for Astrobiology (INTA/CSIC, Madrid), in 2009 he joined the University of Alcalá with the Ramón y Cajal programme. Since 2013 he has been a professor in the area of Chemical Engineering. His scientific activity focuses on environmental biotechnology; specifically, in the bacteria-electrode relationship, pillars of his research, and its application to fields such as wastewater treatment or decontamination of polluted soils, among others. He has supervised 6 doctoral theses, with 12 in preparation, and more than 25 master's theses. He is the author of some 60 scientific publications, 10 book chapters, and 7 national and international patents. He has participated and coordinated 30 national and European projects, including MIDES H2020 as Scientific Manager. He is the scientific director of METfilter, a company owned by IMDEA Agua. In 2016 he received the First Prize of the Intelligent Energy International Campus of Excellence for the project 'Electrosynthesis of bioproducts with high added value and bioenergy from wastewater using mixed cultures of photobacteria illuminated with infrared light'.

About the Book

This book presents the information generated throughout the EU funded MIDES project and includes the latest developments related to desalination of seawater and brackish water by applying an innovative novel technology called microbial desalination cells.

The MIDES project received funding from the European Union's Horizon 2020 research and innovation programme under grant agreement No. 685793. The project was implemented between 2016 and 2020 and aimed to revolutionize desalination by developing a sustainable low-energy process of producing safe drinking water, using microbial desalination cells (MDCs).

The project comprised an international consortium of 10 companies and research organizations (see Table 1) from 7 countries: Austria, Germany, Hungary, the Netherlands, Portugal, Spain and Tunisia. Led by FCC Aqualia, the consortium combined key areas of expertise in water and technology innovation for success of the project.

The project was subdivided into the following nine distinct work packages (WPs), that are designed to lead to the development of the MDC technology and demonstration at designated sites:

- WP1 - Electrodes development
- WP2 - Membrane development
- WP3 - Microbial desalination cell design & bioengineering assays
- WP4 - Process integration and pre-pilot validation

Table 1. List of MIDES partners.

Participant Name	Country	Role
1. Aqualia	Spain	PL
2. Fujifilm	Netherlands	PP
3. IMDEA Water	Spain	PP
4. SGL Carbon	Germany	PP
5. Leitat Technological Center	Spain	PP
6. Mikrolin	Hungary	PP
7. Oncontrol Technologies	Portugal	PP
8. SimTech	Austria	PP
9. IHE Delft Institute for Water Education	Netherlands	PP
10. University of Gabes	Tunisia	PP

PL = Project leader; PP = Project partner.

- WP5 - Process simulation & analysis, automation and control
- WP6 - Pilot plant design and construction
- WP7 - Pilot plant operation and long-term validation
- WP8 - Dissemination & exploitation
- WP9 - Coordination & management

The most relevant results achieved in the 48 (+ 6) months of the project will be described in the following chapters.

Acknowledgements

All MIDES participants wish to acknowledge and to thank the European Commission for approving and supporting the project, in particular Dr. George Kotsikos and Dr. Esther Barrutia, serving as Project Officer and Project Manager, respectively, of the MIDES project in Brussels during the last 4 years.

Chapter 1

Introduction to desalination and microbial desalination cells

Sergio Genaro Salinas-Rodríguez[1], Juan Arévalo[2], Juan Manuel Ortiz[3], Ángeles Mendoza-Sammet[1], Eduard Borràs-Camps[4], Victor Monsalvo-Garcia[2] and Maria D. Kennedy[1]

[1]IHE Delft Institute for Water Education, Water Supply, Sanitation and Environmental Technology Department, Westvest 7, 2611 AX, Delft, Netherlands
[2]Aqualia, Av. del Camino de Santiago, 40, edificio 3, 4ª planta, 28050, Madrid, Spain
[3]Imdea Water, Av/ Punto Com, 2, Parque Científico Tecnológico de la Universidad de Alcalá, 28805, Alcalá de Henares, Madrid, Spain
[4]Leitat, Barcelona, Spain

ABSTRACT

This chapter presents desalination as one of the technologies to alleviate water scarcity and its contribution to the sustainable development goals. An overview of the world and regional desalination capacity is presented and areas where desalination has potential for development are identified. The overall concept of the microbial desalination cells is presented and the areas where key innovations were developed is presented. A discussion on the energy costs and production costs in seawater reverse osmosis is briefly discussed.

Keywords: Brackish water, desalination, MDC, microbial desalination cells, reverse osmosis, seawater, technologies

doi: 10.2166/9781789062120_0001

1.1 INTRODUCTION

Shortage of fresh water is one of the major challenges for societies all over the world. Water desalination can significantly increase the freshwater supply for drinking, industrial use and irrigation. All current desalination technologies require significant electrical or thermal energy, with reverse osmosis (RO) consuming electric energy of at least 3 kWh/m^3 – in extensive tests about 10 years ago, the Affordable Desalination Collaboration (ADC) in California measured 1.6 kWh/m^3 for RO power consumption on the best commercially available membranes, and total plant energy about twice as high, once pre-treatment and pumping is factored in (MacHarg *et al.*, 2008).

To overcome thermodynamic limitations of RO, which points to 1.09 kWh/m^3 for seawater at 50% recovery, microbial desalination cells (MDCs) concurrently treat wastewater (WW) and generate energy to achieve desalination. MDCs can produce around 1.8 kWh of bioelectricity from the handling of 1 m^3 of WW (Gude *et al.*, 2013). Such energy can be directly used to (*i*) totally remove the salt content in seawater without external energy input, or (*ii*) partially reduce the salinity to lower substantially the amount of energy for a subsequent desalination treatment.

The EU-H2020 MIDES project aimed to develop the world's largest demonstrator of an innovative and low-energy technology for drinking water production, using MDC technology as a pre-treatment step for RO.

The project focused on overcoming the current limitations of MDC technology such as low desalination rate, high manufacturing cost, biofouling and scaling problems on membranes, optimization of the microbial-electrochemical process, integration with RO, ceramic nano-membranes as pre-treatment, system scale up and economic feasibility of the technology. This was achieved via innovation in nanostructured electrodes, antifouling membranes (using nanoparticles with biocide activity), electrochemical reactor design and optimization, microbial electrochemistry and physiology expertise, and process engineering and control, such as optimized pre- and post-treatment.

1.2 GLOBAL DESALINATION CAPACITY

Currently, about 21 000 desalination plants are operational with a production capacity larger than 100 Mm^3/d located all over the world in 180 countries. The existing production of drinking water based on RO from various raw water sources is as follows: from seawater 24 Mm^3/d, from brackish water 9 Mm^3/d and from fresh water about 3.2 Mm^3/d. This production of drinking water means that 330 million people in the world receive drinking water supplied by desalination plants (estimated at 120 L/p/d), which is a great contribution to the sustainable development goals, in particular SDG6.

With coastal population densities on the increase in many parts of the world – and considering that nearly 2.4 billion people live within 100 km of the coast (UN,

Ocean Conference 2017) – the development potential for seawater desalination facilities is huge. Brackish water treatment and water reuse also offer a great future potential for alleviating water stress, in particular for industrial applications.

Figure 1.1 presents the global historical cumulative production capacity of desalination plants for all raw water sources, including: seawater, brackish water, fresh water, wastewater, and pure water. Over two-thirds of the current total capacity is produced by membrane-based desalination technology (reverse osmosis) and less than one-third is produced by thermal processes (multi-stage flash distillation [MSF], and multi-effect distillation [MED]). One of the reasons why seawater reverse osmosis production capacity is growing faster than thermal processes is the lower investment costs and the lower energy consumption (3–4 kWh/m^3). In the last 30 years, the online production capacity has increased from 13.7 Mm^3/d to the current 101.6 Mm^3/d, which is about 7.5× more capacity. In the last 10 years, the growth in desalination capacity has been about 41% and mostly related to the new plants making use of reverse osmosis as the main desalination technology. It is expected that the production capacity of desalination plants by 2030 will double the current capacity.

The implementation of desalination plants has increased in many parts of the world. Much of the growth of the desalination capacity takes place in the seawater industry, although wastewater desalination and brackish water desalination is becoming more relevant. In many countries, such as the Netherlands, reverse osmosis is replacing conventional groundwater treatment, due to the robust approach to also remove micropollutants (endocrine disruptors, medicines, personal care products, micro-plastics, etc.) that could be present in the raw water sources.

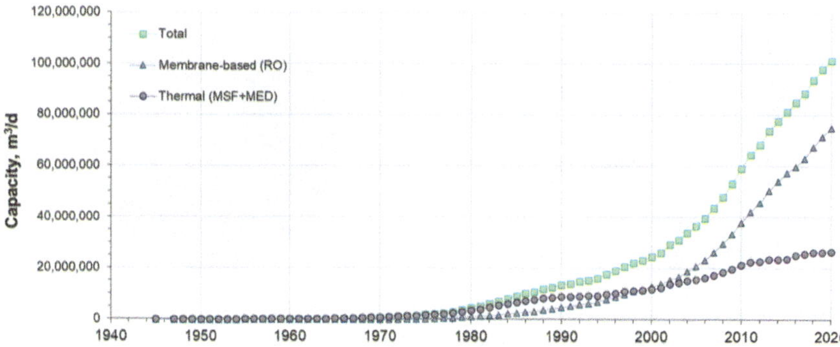

Figure 1.1 Total desalination capacity in the world (seawater, brackish water, wastewater, and fresh water). Source: Global Water Intelligence (2020) 32nd Worldwide Desalting Plant Inventory Media Analytics Ltd.

1.2.1 Capacity by desalination technology and source water type

In 2020, 8% of the total capacity of 101.6 Mm^3/d was produced from wastewater; 20% of the capacity was produced from brackish water sources, mainly brackish groundwater, and ~60% of the capacity was produced from seawater (see Figure 1.2). Seawater is hence the predominant source water for desalination and accounts for a worldwide water production of ~60 Mm^3/d.

For all source water types reverse osmosis (RO) is the preferred desalination technology. It accounts for about 70% (67 Mm^3/d) of the global capacity (Figure 1.2); 24% or 23.2 Mm^3/d of the global capacity is produced by distillation plants, either multi-stage flash (MSF) or multi-effect distillation (MED) plants, with relative market shares of 17% (16.6 Mm^3/d) and 7% (6.6 Mm^3/d), respectively. The electrodialysis (ED) process has a market share of around 2% (1.97 Mm^3/d). Other processes used, include electro-di-ionization (EDI) accounting for 0.3% (0.3 Mm^3/d) and nano-filtration (NF) accounting for another ~2% (1.8 Mm^3/d) of the world desalination capacity.

The picture changes if one distinguishes between the different source-water types (see Figure 1.3). For seawater, (seawater reverse osmosis) SWRO and thermal processes dominate global sea water desalination production (34.4 and 25.7 Mm^3/d, respectively). MSF is the main thermal process, accounting for 31% of the global seawater desalination production. On the contrary, RO is the dominant process for brackish water (90%, 17.8 Mm^3/d) and wastewater (91%, 6.9 Mm^3/d) desalination, whereas distillation plays only a minor role for these source waters.

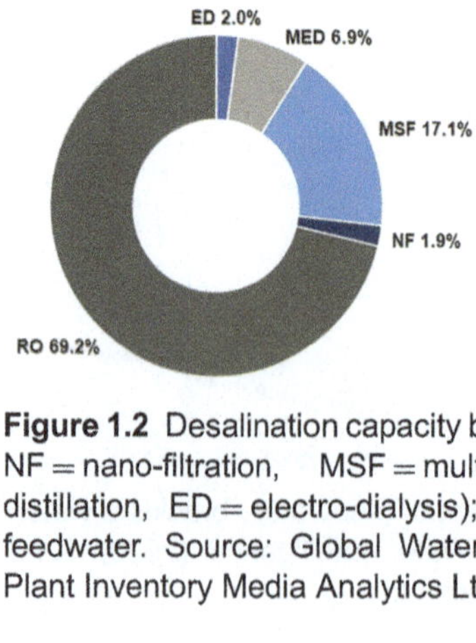

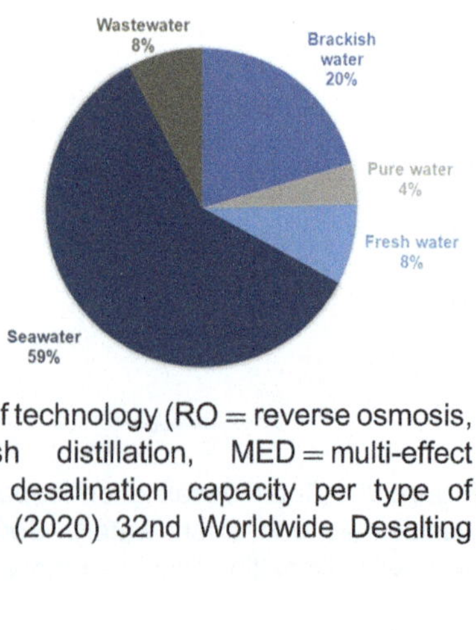

Figure 1.2 Desalination capacity by (left) type of technology (RO = reverse osmosis, NF = nano-filtration, MSF = multi-stage flash distillation, MED = multi-effect distillation, ED = electro-dialysis); and (right) desalination capacity per type of feedwater. Source: Global Water Intelligence (2020) 32nd Worldwide Desalting Plant Inventory Media Analytics Ltd.

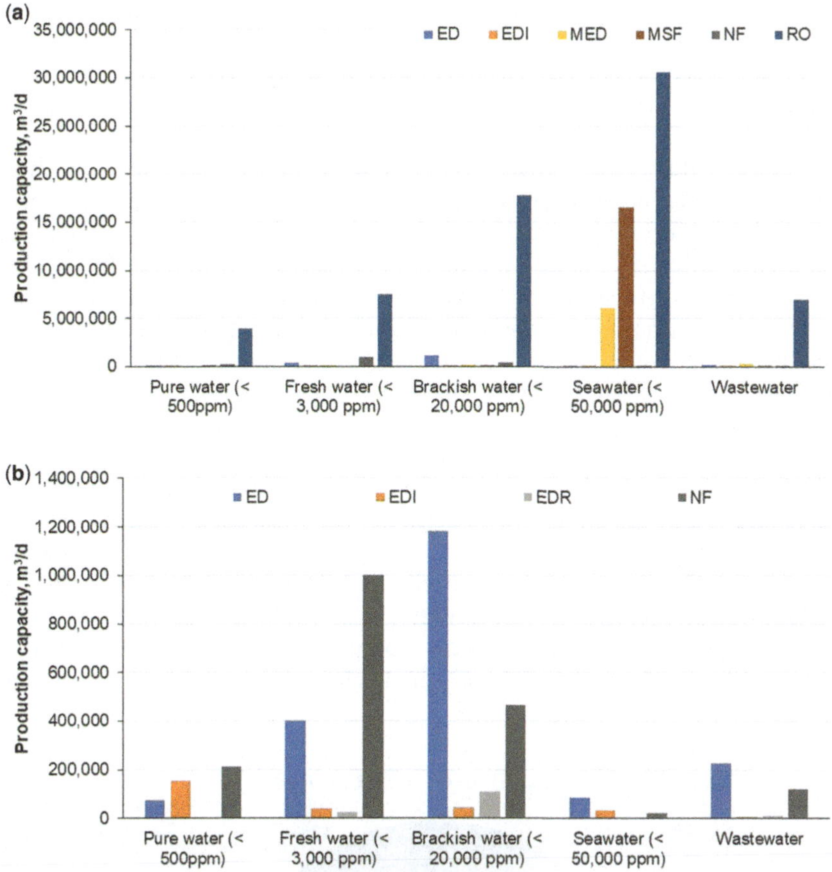

Figure 1.3 Desalination production capacity in m^3/d per raw water source and per technology for plants online and presumed online. ED = electrodialysis, EDI = electro-de-ionization, MED = multi-effect distillation, MSF = multi stage flash, NF = nanofiltration, RO = reverse osmosis, EDR = Electrodialysis reversal. Source: Global Water Intelligence (2020) 32nd Worldwide Desalting Plant Inventory Media Analytics Ltd.

1.2.2 Desalination capacity by region and source water type

In 2020, approximately 53% (43.9 Mm^3/d) of the global desalination production takes place in the Middle East and North Africa. In addition, ~16% (16.1 Mm^3/d) of the global capacity is located in East Asia and pacific regions, 10% (10 Mm^3/d) in North America, 8% (8 Mm^3/d) in Western Europe, 6% (5.7 Mm^3/d) in Latin America and the Caribbean, 2% (1.8 Mm^3/d) in sub-Saharan Africa, 3% (3.3 Mm^3/d) in South Asia, and 2% (2.2 Mm^3/d) in Eastern Europe and Central Asia.

Figure 1.4 shows the percentage desalination capacity per region considering exclusively seawater as the source, where the Middle East and North Africa accounts for 69%.

As illustrated in Figure 1.5, in most of the world regions, seawater is the main water source for desalination with the exception of North America where brackish water desalination accounts for 73% (7.3 Mm^3/d) of the regional capacity followed by 19% wastewater (1.9 Mm^3/d). Seawater desalination accounts for 79% of the capacity in Latin America and the Caribbean, compared to sub-Saharan Africa (77%), the Middle East and North Africa (84%), Southern Asia (61%), and Western Europe (65%); and is the predominant process in most remaining regions except for North America; Northern Europe; and Japan, Korea, and Taiwan.

Japan, Korea, and Taiwan's combined desalination production of about 1.95 Mm^3/d is produced from seawater (35%), brackish water (29%) and wastewater (36%). In the case of Singapore, the production capacity is about 2 Mm^3/d produced from seawater (55%), brackish water (2%) and wastewater (43%). Australia has a production capacity of 2.9 Mm^3/d from seawater (63%), brackish water (15%) and wastewater (22%).

The countries with desalination capacities larger than 650 000 m^3/d are presented in Figure 1.6. The use of desalination for irrigation is relevant

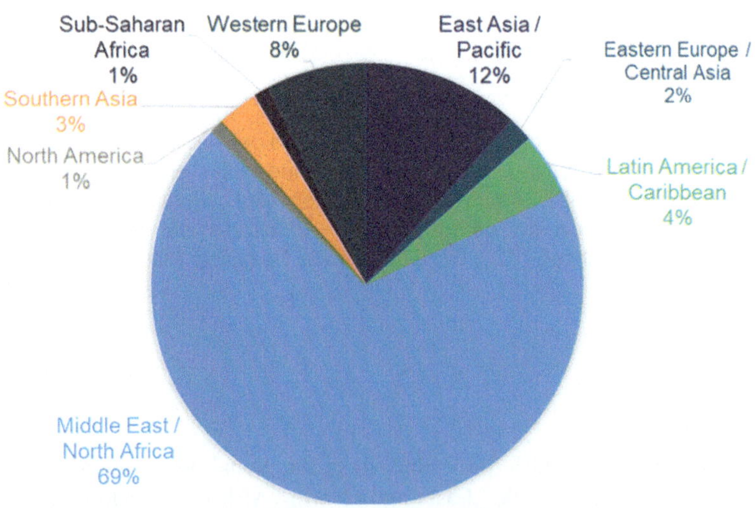

Figure 1.4 Desalination per sea water reverse osmosis location. Source: Global Water Intelligence (2020) 32nd Worldwide Desalting Plant Inventory Media Analytics Ltd.

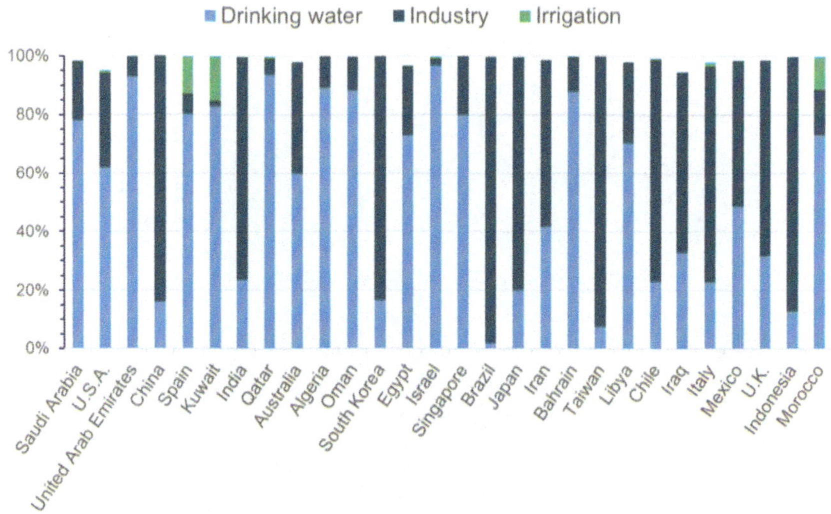

Figure 1.5 Highest capacity desalination countries and customer type. Source: Global Water Intelligence (2020) 32nd Worldwide Desalting Plant Inventory Media Analytics Ltd.

in three countries, namely, Spain, Kuwait, and Morocco. China, India, South Korea, Brazil, Japan, Taiwan, and Indonesia rely on desalination for industry applications. Saudi Arabia, USA, UAE, Spain, Kuwait, Algeria, Oman, Israel, Singapore, Bahrain, Libya, and Morocco rely on desalination for municipal use.

In conclusion, about 68% or 68.5 Mm^3/d of the worldwide desalination capacity was produced from seawater sources in 2020. The global desalination capacity increased by 41% compared to the year 2010 (59.2 Mm^3/d). Of the desalinated seawater, 57% is produced by reverse osmosis. The MSF distillation process is reserved almost exclusively for the desalination of seawater, mainly in the Gulf countries.

1.3 ABOUT THE MIDES PROJECT

The MIDES project aimed to revolutionize desalination by developing a sustainable low-energy process of producing safe drinking water, using Microbial Desalination Cells (MDCs) as a pre-treatment for reverse osmosis (RO). The project ran from 2016 to 2020, to develop the world's largest demonstrator of the innovative MDC technology. Demonstration sites were implemented in Spain in Denia (brackish water) and in Tenerife (seawater).

The MIDES project comprised an international consortium of 10 companies and research organizations from seven countries: Austria, Germany, Hungary, the Netherlands, Portugal, Spain and Tunisia.

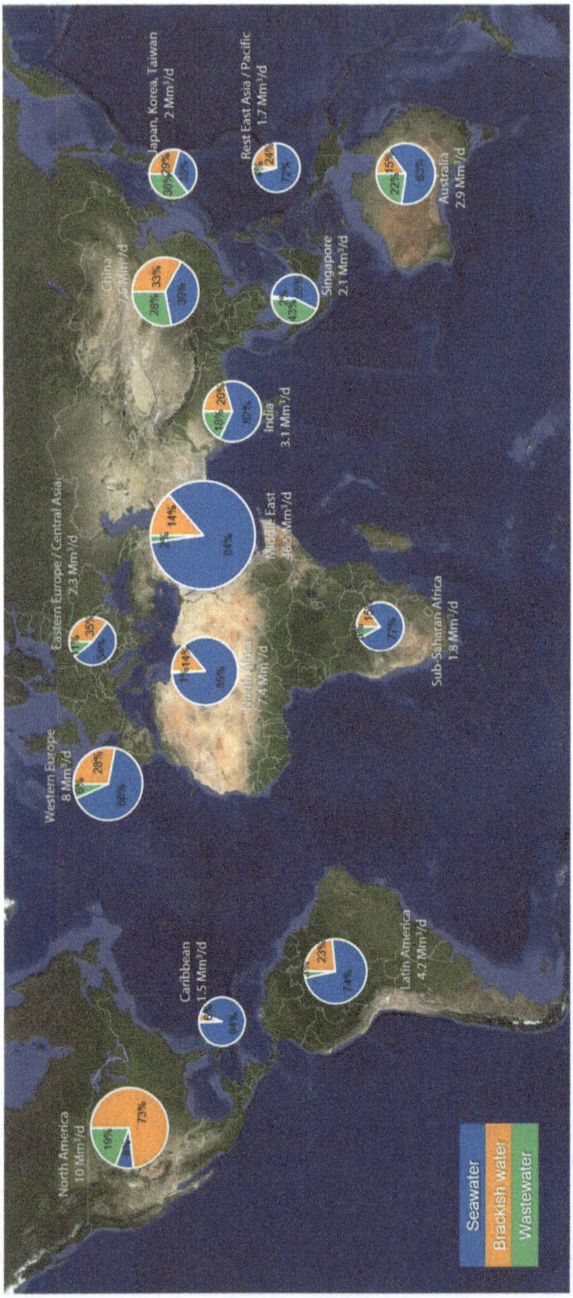

Figure 1.6 Desalination capacity in different regions in the world per percentage capacity production of various water sources (seawater, brackish water, wastewater). For example: North America desalinates water with a total capacity of 10 Mm^3/d of which 73% is produced from brackish water, 19% from wastewater and 8% from seawater. Source: Global Water Intelligence (2020) 32nd Worldwide Desalting Plant Inventory Media Analytics Ltd.

1.4 OVERALL MIDES CONCEPT

As shown in Figure 1.2, the desalination market is mostly dominated by reverse osmosis (RO), compared to competition between multi-stage flash (MSF) and multi-effect distillation (MED). However, the high energy cost continues to be a major concern, with energy consumption accounting for 75% of the desalination operating cost when excluding capital costs or 40% including capital costs (ElMekawy *et al.*, 2014). This energy cost for desalination is about 10 times higher than for conventional water sources, leading to high water prices, that can easily exceed 0.5 €/m^3.

Recently, a new technology called microbial desalination cell (MDC) was shown to desalinate water without the use of any external electrical power. Electrical energy produced directly by the degradation of organic matter in water by bacteria was used to achieve desalination. Derived from microbial fuel cells (MFCs), MDCs use electric potential generated from the microbial metabolism of organic compounds to drive desalination, similar to electrodialysis (ED). The advantages of an MDC include less external energy for the desalination process and simultaneous wastewater treatment (WWT). Moreover, MDCs operate under neutral pH, pressure and temperature conditions (Zhang & He, 2012).

A typical MDC consists of three chambers: an anode, middle (salt), and a cathode (Figure 1.7). These are separated by an anion exchange membrane (AEM, between

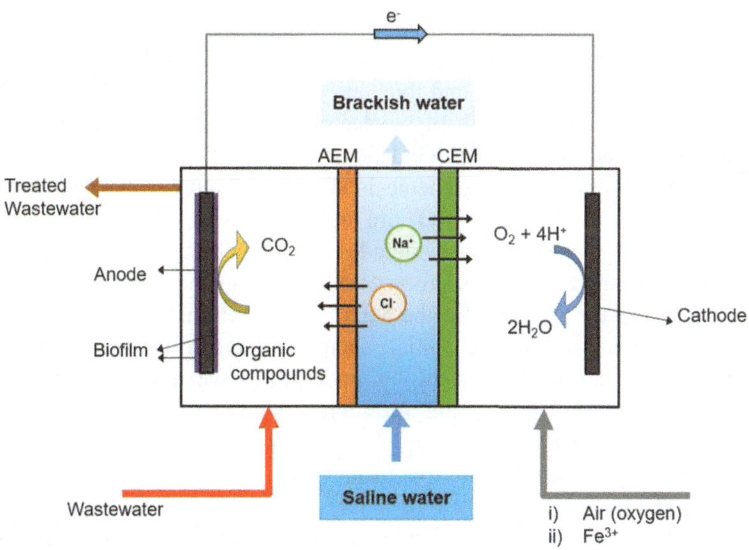

Figure 1.7 The schematic of a microbial desalination cell (MDC) as applied in the MIDES concept. AEM: anion exchange membrane; CEM: cation exchange membrane. Bioelectrogenic biofilm on the anode provides electric potential for ion migration. Source: Figure adapted from Cao *et al.* (2009).

the anode and the middle chambers) and a cation exchange membrane (CEM, between the cathode and the middle chambers) (Saeed *et al.*, 2015). Bacteria growing on the carbon-based anode electrode break down organic matters while releasing electrons and protons in the anode chamber. Terminal electron acceptors (e.g., ferricyanide or oxygen) in the cathode chamber are reduced by accepting those electrons through an external circuit. Anions and cations are driven to migrate through the AEM and CEM to balance the electric charge in the anode and the cathode chambers. Consequently, the water in the middle chamber is desalinated by ion migration from the central compartment to the adjacent ones.

In 2009, the first lab-scale MDC devices were able to provide only a few millilitres of drinking water (Cao *et al.*, 2009), whereas the desalinated water production in the largest scaled-up MDC before the start of the MIDES project was reported to produce 0.077 $L/m^2/h$ (for partial desalination from 52.4 to 20 mS/cm) (Zhang & He, 2015). While complete desalination in MDCs can be achieved without an electrical external source, the specific production of such a 'microbial-passive' system is still 200 times lower compared to conventional desalination systems (for RO = 15–20 $L/m^2/h$).

To increase the desalinated water production while maintaining low energy requirements, one of the options explored in MIDES is MDC technology as pre-desalination in connection with conventional RO desalination units. Since the RO process efficiency and energy requirement largely depends on feed water salinity, in which salt permeation through the membrane increases proportionally with the salt concentration of the feed water, pre-treatment of the feed water with an MDC will reduce its salinity and thus energy demands for downstream RO. It is predicted that around 1.8 kWh of bioelectricity can be produced by an MDC from the handling of 1 m^3 of WW (Gude *et al.*, 2013).

The MIDES overall process scheme is shown in Figure 1.8 where the core technologies, named MDC and RO, are integrated with other complementary technologies. The overall process includes the initial treatment of municipal wastewater in a conventional anaerobic reactor and in the meantime using the effluent of the reactor as a fuel for the MDC, which returns a treated wastewater with a 90% reduction of the initial chemical oxygen demand (COD). In conventional RO desalination, seawater undergoes several pre-treatment steps (chemical coagulation, settling and filtration) to protect the membranes from particles and organic matter.

In MIDES, this whole pre-treatment was substituted by nano-coated ceramic membranes. This resulted in a relevant reduction in chemicals usage and footprint as well as lowering the energy demand by 80% (going from ~0.48 to <0.1 kWh/m^3). After pre-treatment, seawater entered the MDC unit where it was partially desalinated (70–90%) before it was fully processed in the RO unit.

The work on MIDES was based on two main actions:

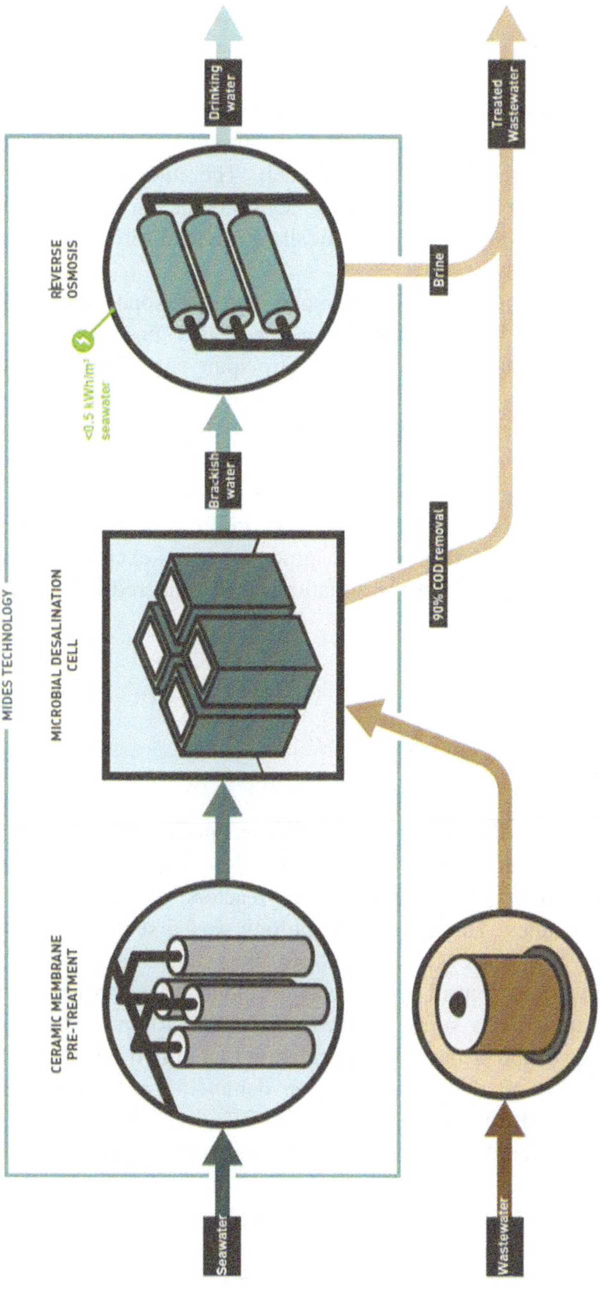

Figure 1.8 MIDES concept scheme. Source: Own creation.

1.4.1 MDC process optimization

The performance of MDCs can be improved by overcoming the main barriers of the technology. MDCs were tested in a relevant environment for both wastewater treatment and saline water desalination aiming at a salt removal rate of 75–90% before RO and a COD removal of 90%. The microbial conversion of organic matter into electric current was maximized by a microbial-electrochemical engineering approach. The microbial activity in the anodic chamber was promoted by using salt-tolerant mixed populations of bacteria from the *Geobacter* genus, specifically generated through an innovative gradostat selection process.

High performance electrodes and membranes were developed and implemented into the MDC, using large surface area electrodes based on 3D porous nanostructured carbon materials such as electro-spun nanofibres, graphene and carbon nanotubes (with emphasis on mechanically strong, cheap and easily scalable materials). This also included the coating or impregnation of commercial low-cost porous electrodes with highly electro-active nano-additives.

Innovative anti-fouling membranes were fabricated by implementing biocidal activity nano-coatings (biocidal nanoparticles, grafting of functional groups, layer-by-layer) into the current production process for ion-exchange membranes. The final approach to improve the desalination rate considered innovation in cell design and engineering.

1.4.2 MDC-RO technology validation, up-scaling and testing

The integration of the MDC process with RO technology was first studied by simulation and modelling. Then, an optimized cell design and components were employed to construct two pre-pilot MDC reactors, to be integrated with commercial RO modules. This hybrid technology was first tested and validated with real seawater and wastewater. To ensure the achievement of the energetic goals, an automation and control system was developed, featuring a smart energy management tool specifically suited for the hybrid MDC-RO process.

The crucial stage of the project was the construction of two full-scale MDC-RO demonstrators, implemented in representative demo-sites in Spain (Denia, and Tenerife) and operated for several months (150 L/h of drinking water with energy consumption below 0.5 kWh/m^3 accompanied with WW treatment capacity of 1–3 m^3/day).

Finally, life cycle assessment (LCA) was applied as a tool to analyse the environmental impact of the MIDES Project technology along its whole life cycle including fabrication, use and disposal. Besides this a full risk assessment was performed.

1.5 SUMMARY

MDCs are based on bio-electro-chemical technology, in which biological wastewater treatment can be coupled to the desalination of a saline stream using ion exchange membranes without external energy input. MDCs simultaneously treat wastewater and perform desalination using the energy contained in the wastewater. In fact, an MDC can produce around 1.8 kWh of bioelectricity from the energy contained in 1 m^3 of wastewater. Compared to traditional RO, more than 3 kWh/m^3 of electrical energy is saved. With this novel technology, two low-quality water streams (saline stream, wastewater) are transformed into two high-quality streams (desalinated water, treated wastewater) suitable for further uses.

An exhaustive scaling-up process was carried out and several innovations were produced related to nanostructured electrodes, antifouling membranes, electrochemical reactor design and optimization, microbial electrochemistry and physiology expertise, and process engineering and control.

Nominal desalination rates of 4–11 $Lm^{-2}\,h^{-1}$ were reached with a current efficiency of 40%. After the scalability success, two MDC pilot plants were designed and constructed consisting of one stack of 15 MDC pilot units with a 0.4 m^2 electrode area per unit.

This configuration is being validated at pilot scale at two demonstration sites to prove the functionality and competitiveness of the system in different environments and with different water sources: brackish water and seawater. All components of the pilot plant designed in the MIDES project are located at the first demonstration site in Denia, Spain: brackish water pre-treatment, wastewater pre-treatment, the MDC, low-pressure RO, and post-treatment including remineralization and disinfection. This working configuration has been validated by treating brackish water from the Racons River to produce safe drinking water according to national and European regulations, at a very low energy and with positive environmental benefits.

The second demonstration site is at SWRO plant Oeste, in Guia de Isora (Tenerife, Spain). Partial desalination of seawater is performed by the MDC, without external energy, for its subsequent post-treatment using low pressure RO. Desalination is powered by chemical energy contained in residual organics, effectively transformed in electric current thanks to the bioelectrogenic activity of the microbial consortia growth in the isolated wastewater compartment.

More sustainable desalination can be achieved with MDC technology. For example, with the potential savings generated with the MDC, in a city of 10 000 inhabitants, 2500 t CO_2/y can be saved using MDC technology instead of conventional treatment systems. This is roughly equal to the same impact as planting 115 500 trees per year. The MDC process can also be adapted for water/energy production. If desalinated water is not needed at some point, it is possible to generate electricity, and the wastewater treatment plant can provide up to 7.2% of the energy needs of the wastewater population served.

REFERENCES

Cao X., Huang X., Liang P., Xiao K., Zhou Y., Zhang X. and Logan B. E. (2009). A new method for water desalination using microbial desalination cells. *Environmental Science & Technology*, **43**, 7148–7152, doi: 10.1021/es901950j

ElMekawy A., Hegab H. M. and Pant D. (2014). The near-future integration of microbial desalination cells with reverse osmosis technology. *Energy & Environmental Science*, **7**, 3921–3933, doi: 10.1039/C4EE02208D

Global Water Intelligence. (2020). 32nd Worldwide Desalting Plant Inventory Media Analytics Ltd.

Gude V. G., Kobakian B. and Gadhamshetty V. (2013). Beneficial bioelectrochemical systems for energy, water, and biomass production. *Journal Microbial Biochemical Technology*, **S6**, 5948

MacHarg J., Seacord T. F. and Sessions B. (2008). Affordable desalination collaboration (ADC) baseline tests reveal trends in membrane performance. *Desalination and Water Reuse*, **18**, 30–39

Saeed H. M., Husseini G. A., Yousef S., Saif J., Al-Asheh S., Abu Fara A., Azzam S., Khawaga R. and Aidan A. (2015). Microbial desalination cell technology: A review and a case study. *Desalination*, **359**, 1–13, https://doi.org/10.1016/j.desal.2014.12.024 (Accessed: 30 March 2021)

Zhang B. and He Z. (2012). Energy production, use and saving in a bioelectrochemical desalination system. *RSC Advances*, **2**, 10673–10679, doi: 10.1039/C2RA21779A

Zhang F. and He Z. (2015). Scaling up microbial desalination cell system with a post-aerobic process for simultaneous wastewater treatment and seawater desalination. *Desalination*, **360**, 28–34, https://doi.org/10.1016/j.desal.2015.01.009 (Accessed: 30 March 2021)

Chapter 2

Microbial desalination cell design & bioengineering assays: Main concepts

Marina Ramírez-Moreno[1], Pau Rodenas[1], Martí Aliaguilla[2], Pau Bosch-Jimenez[2], Eduard Borràs[2], Patricia Zamora[3], Víctor Monsalvo[3], Frank Rogalla[3], Juan M. Ortiz[1] and Abraham Esteve-Núñez[1,4]

[1]IMDEA Water Institute, Avenida Punto Com, 2, Parque Científico Tecnológico de la Universidad de Alcalá, 28805, Alcalá de Henares, Madrid, Spain
[2]LEITAT Technological Center, C/Pallars, 179-185, 08005, Barcelona, Spain
[3]FCC Aqualia S.A., Avenida del Camino de Santiago, 40, Building 3, 4th floor, 28050, Madrid, Spain
[4]Department of Analytical Chemistry, Physical Chemistry and Chemical Engineering Department, Universidad de Alcalá, Alcalá de Henares, Spain

ABSTRACT

The main goal of this chapter is to present the main concepts and principles for the microbial desalination process. Also, a rational explanation of the electrochemical behaviour of the microbial desalination cell (MDC) setup under different experimental conditions is presented. The final section of the chapter shows the design and construction of an MDC pre-pilot unit, as well as the main results of desalination and water treatment capacity for the scaled-up device.

doi: 10.2166/9781789062120_0015

Keywords: Brackish water, desalination, microbial electrochemical technology, seawater, wastewater treatment

2.1 PRINCIPLES, MAIN CONCEPTS AND STATE OF THE ART

Microbial electrochemical technologies (METs) constitute a platform (Schröder *et al.*, 2015) of emerging technologies based on the interaction between electroactive bacteria and electrodes. Indeed, a variant of the fuel cell, the microbial fuel cell (MFC), allows the direct transformation of soluble organic matter into electric current (Borjas *et al.*, 2015). The produced power (in the range of 0.001–19 W m^{-2}, referenced to cross section) (Logan *et al.*, 2015) could be used in the same system to enhance organic matter degradation and drive other processes requiring electrochemical energy, and operate in a self-sufficient and decentralized fashion as a passive system.

In this microbial electrochemical context, the microbial desalination cell (MDC) could be considered as the integration of an MFC and an electrodialysis (ED) cell in order to treat wastewater and desalinate seawater. By using the energy provided by the oxidation of organic matter, contained in the wastewater, this system drives the migration of ions and the desalination process. Thus, MDC technology is able to desalinate saline water without consuming electric or thermal energy and allowing the use of the energy for any other processes. In this sense, MDC technology could be employed to save energy and avoid the greenhouse gases related to conventional processes (seawater RO produces 1.78 kg of CO_2 per m^3 using 600 g CO_2 kWh^{-1} in the average European Union (EU) energy mix). Indeed, the versatile and simultaneous applications of MDC have the potential to be a real and feasible alternative for both desalination and wastewater treatment (Cao *et al.*, 2009).

The MDC unit is composed of at least three chambers (Figure 2.1): (1) an anaerobic anodic chamber that contains the electroactive microbial community which first oxidizes the organic matter (fuel) contained in the wastewater and then transfers electrons to the anode; (2) a central desalination compartment separated from the others by an anion exchange membrane (AEM) and a cation exchange membrane (CEM); and (3) a cathodic chamber where the reduction counter-reaction occurs. In an MDC reactor, CEM and AEM are alternatively placed between the cathodic and the anodic compartment, as indicated in Figure 2.1. Moreover, an external load is placed between the anode and cathode collector, allowing the flow of electric current. If organic matter is used to feed the anodic compartment, and the cathodic compartment is fed with a catholyte (e.g., an Fe^{3+} complex or oxygen in acidic solution), then a potential difference is established between the electrodes. Due to the differential charges of the two chambers, anions and cations migrate through the respective membranes raising

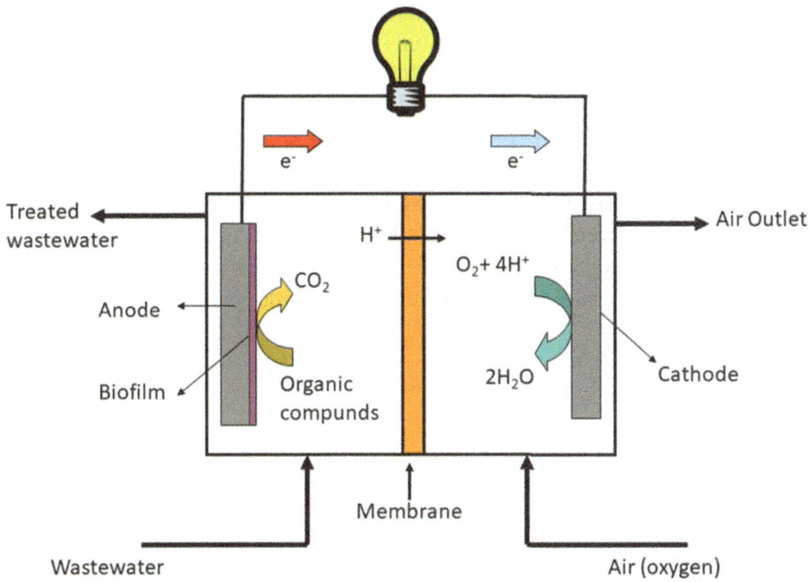

Figure 2.1 The schematic of a microbial desalination cell (MDC). AEM: anion exchange membrane; CEM: cation exchange membrane. Bioelectrogenic biofilm on the anode provides electric potential for ion migration. *Reprinted from Strategies for merging microbial fuel cell technologies in water desalination processes: Start-up protocol and desalination efficiency assessment, Vol. 356, Pages 519–528, 2017, with permission from Elsevier.*

the salt concentration in the anodic and cathodic compartments while decreasing it in the saline compartment (central compartment). Thus, the MDC constitutes a sustainable process since it does not require external energy input because microbes harvest it from waste.

From the first proof-of-concept proposed by Cao (2009), different MDC designs have been reported in the literature, including cubic and tubular reactors (Jacobson *et al.*, 2011a; b; Mehanna *et al.*, 2010; Ping *et al.*, 2013), stacked cells (Chen *et al.*, 2011; Kim & Logan, 2011a), using batch recirculation (Chen *et al.*, 2012; Qu *et al.*, 2012), biocathode MDCs (Wen *et al.*, 2012), increasing water production by applying external voltage (Ge *et al.*, 2014), or integrating innovative membranes (forward osmosis) (Yuan *et al.*, 2015; Zhang & He, 2012), ion exchange resins in the compartments (Zhang *et al.*, 2012) or microfiltration processes (Zuo *et al.*, 2017, 2018). Although most of the MDC prototypes studied in the literature are at lab scale (i.e., millilitres), a 100 L pilot-scale MDC unit achieving partial desalination of seawater has been reported in the literature (Zhang & He, 2015). Thus, MDC technology has been proposed in the literature as a suitable pre-treatment for reverse osmosis to reduce energy costs for desalination

(ElMekawy *et al.*, 2014), and indeed this concept is aligned with the MIDES process, as indicated in Chapter 1, section 1.4: 'Overall MIDES Concept'.

Regarding the performance of the MDC, among the main limitations are (i) the drastic increase of the internal resistance due to the changes in the conductivity of the saline medium, and (ii) the pH variations in the anodic chamber that can affect the biofilm growth and performance. Related to the efficiency of the desalination process, the main constraints are limited potential to drive ion migration, back diffusion of salts between anodic/cathodic and saline compartments as well as water trespassing though membranes. Understanding the limitations at lab scale could help for further scale-up of the technology at pilot-plant scale.

2.2 MDC INLET STREAM: STUDY OF PRE-DIGESTION OF WASTEWATER

An MDC uses energy provided by the oxidation of organic matter contained in the wastewater, however the feed must be pre-treated in order to obtain a highly-enriched inlet with volatile fatty acids (rapidly degradable organic matter). Thus, anaerobic pre-digestion can be seen as a key element for MDC systems, since it provides the energy required for the desalination step.

This section summarizes experimental results from several hydrolytic acidogenic anaerobic bioreactors, which were designed, built, started-up and operated with the objective of producing an enriched acetate effluent to be fed into the MDC. With this pre-digested fuel (i.e., high concentration of acetate), the anode associated biofilm can generate electric current in an efficient manner. The proposed design for the anaerobic bioreactors for this specific application was an upflow anaerobic sludge blanket reactor (UASB). Pre-pilot scale reactors (5.2 and 11 L) have been built according to the following diagram (Figure 2.2):

These reactors were being operated at different hydraulic retention times (HRT) with the aim of obtaining the maximum amount of volatile fatty acids (VFA) and acetate in the effluent stream. The performance of the anaerobic bioreactors was assessed by the biogas production and the conversion of soluble chemical oxygen demand (COD) in the influent to VFA and acetate in the effluent (ratio mass acetate COD (effluent) per mass soluble COD (influent)).

Four anaerobic reactors were operated to increase the VFA concentration and water quality of the raw wastewaters, as described below:

- **Reactor 1**: UASB inoculated with flocculated biomass fed with raw municipal wastewater doped with molasses (1% v/v).
- **Reactor 2**: UASB inoculated with granular biomass fed with raw municipal wastewater doped with molasses (1% v/v).
- **Reactor 3**: UASB reactor fed with raw municipal wastewater.
- **Reactor 4**: EGSB (expanded granular sludge bed) reactor inoculated with granular biomass fed with industrial wastewater (brewery industry).

Table 2.1 shows the obtained results for the four operated anaerobic reactors.

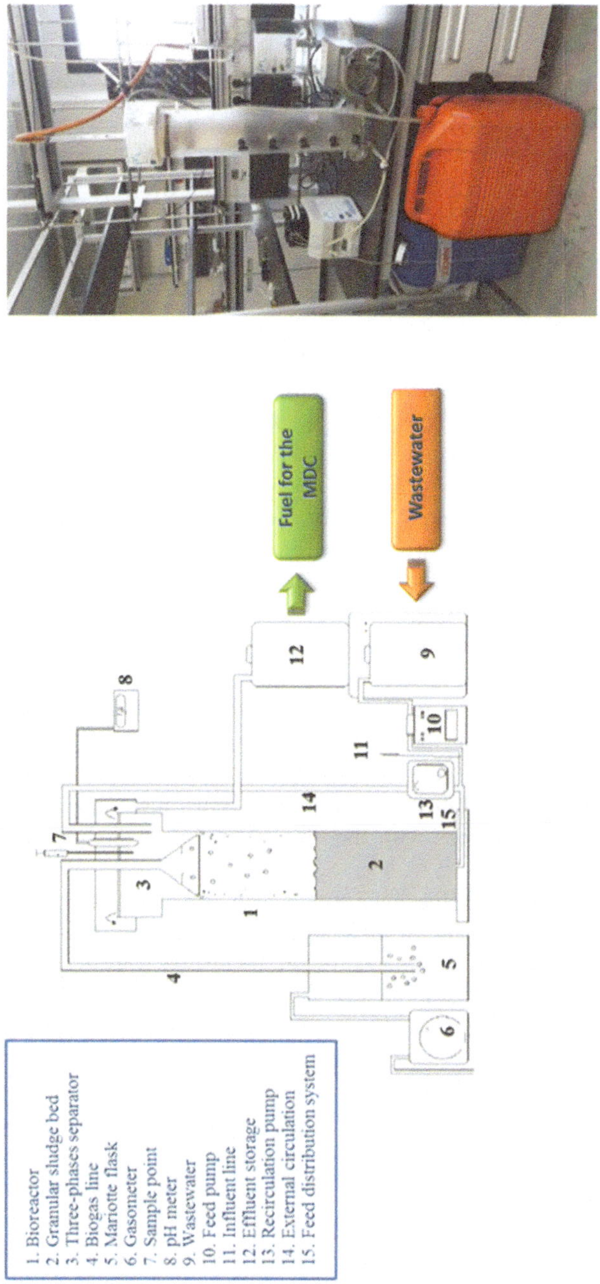

Figure 2.2 High-rate anaerobic reactors for wastewater pre-digestion. Source: Aqualia.

Table 2.1 Average VFA concentration in the anaerobic reactor respective effluents.

	Concentration of VFA (mg L^{-1})						
	Acetic Acid	Propionic Acid	Butyric Acid	Valeric Acid	Isocaproic Acid	Caproic Acid	Total
Reactor 1	1199	521	913	797	180	101	3713
Reactor 2	1432	705	944	1059	97	193	4446
Reactor 3	36	5	8	17	6	6	71
Reactor 4	277	123	4	5	9	5	429

Reactor 1: No effect of the HRT on the VFA production was observed. The results showed a high productivity of VFA sufficient to power an MDC. Acetic acid and propionic acid, two of the VFA with the smaller carbon chain (2 and 3 carbons, respectively), were the main forms detected in the analysis. This confirms the suitability of wastewater plus molasses as a potential feed to be treated in an anaerobic reactor operated with flocculent anaerobic biomass to maximize VFA production.

Reactor 2: The microbiology in Reactor 2 presents a granular configuration, which allows for a proper solid removal in the bioreactor. Additionally, Reactor 2 performed better compared to Reactor 1 in terms of VFA production due to the microbiology present in the granular biomass. This VFA concentration is sufficient to guarantee the feed composition for efficient MDC operation and confirms the suitability of wastewater plus molasses as influent using granular anaerobic reactors with the aim of producing organic acids for MDC operation.

Reactor 3: Anaerobic Reactor 3 was a high-rate anaerobic reactor operated at different HRTs (lower than 5 h) treating urban wastewater. Municipal wastewater showed a much lower COD concentration, compared with the COD observed in the mix of wastewater plus molasses. This COD concentration presented a lower potential of VFA generation, since organic matter is the main precursor for VFA generation.

Reactor 4: Reactor 4 was operated with HRT from 8 to 2 h treating brewery wastewater. This wastewater showed a COD lower than that of Reactor 1 and 2 and higher than that of Reactor 3.

In light of these results, the pre-digested municipal wastewater doped with molasses and the non-digested industrial wastewater could serve as fuel for the MDC given their high VFA content. The latter was selected as the more convenient effluent for future pre-pilot MDC experiments (see section 2.5: Design, construction and operation of MDC pre-pilot unit).

2.3 CATHODE LIMITATIONS FOR MDC: AIR DIFFUSION CATHODE AND LIQUID CATHOLYTE

The first study carried out by Cao (2009) used a ferricyanide catholyte (i.e.$Fe(CN)_6^{3-} + 1e^- \rightarrow Fe(CN)_6^{4-}$, $E^0 = 0.36$ V) to demonstrate the feasibility of the MDC process, just providing enough electric potential in the system to produce the complete desalination (i.e., salt removal 88–94%, energy production 2 W m^{-2}). In subsequent studies reported in the literature (Saeed *et al.*, 2015), the use of oxygen reduction was normally implemented as the cathodic reaction, decreasing both the water production and energy production, and achieving complete desalination using external energy application (i.e., salt removal 3.7 to 9.2 kg m^{-3} using external voltage) (Zhang & He, 2015). Other strategies, such as biocathodes or stacked systems, have been reported in the literature with similar results (i.e., salt removal 44–93% using external voltage) (Kim & Logan, 2011a; Wen *et al.*, 2012).

Even though oxygen reduction is a convenient option for cathode reaction in MDC systems (i.e., low cost, sustainable), it is interesting to study the impact of the use of a redox mediator (such as ferro-ferricyanide) in the desalination performance (i.e., salt removal) and electric energy production, as a better performance has been observed experimentally when used in the cathodic compartment. Obviously, the use of ferro-ferricyanide catholyte (or other redox mediators/compounds) would only be feasible from a technical point of view in MDCs if (i) the redox mediator is low cost or (ii) an easy and cheap strategy is developed for regeneration of catholyte solution once depleted. In all cases, understanding the electrochemical behaviour and the improvement in the performance of the MDC when different cathodic reactions are used could help to increase the water production in MDC technology, one of the main limitations of this sustainable technology for real applications.

Table 2.2 shows a summary of conventional three-chambered MDC studies with synthetic solutions and ferricyanide or air diffusion cathode.

2.4 MDC RESULTS AT LAB SCALE

In these sections, experimental results from two MDCs at lab scale are presented. Bioelectrochemical reactors were constructed and operated at the same experimental conditions (Figure 2.3) to generate comparable experimental results using different cathodic reduction strategies: liquid catholyte (containing ferro-ferricyanide solution 0.06 M as electron acceptor, pH = 7.4) and air diffusion cathode (using atmospheric oxygen as electron acceptor). MDC prototypes were employed for desalination experiments, at IMDEA and LEITAT laboratories, with a compact stack design consisting of several polypropylene compartments and neoprene or rubber gaskets for an optimal hermetic seal. The three compartment configuration was used. The device was closed with stainless

Table 2.2 Summary of conventional three-chambered MDC studies with synthetic solutions and ferricyanide or air diffusion cathode.

Operation Mode	MDC Chamber							Main Parameters		Ref.
	Anodic Chamber			Cathodic Chamber		Desalination Chamber		Desalination	Treatment	
	Electrode		Electron Donor	Electrode	Catholyte	Initial Salinity (g L^{-1} NaCl)	End Desal. Cycle	Salt Removal*	COD Removal	
Batch	Carbon felt 9 cm^2		Acetate 1.6 g L^{-1} 5 mL min^{-1}	C. felt	Ferricyanide + PBS** 5 mL min^{-1}	5	E_{cell} <50 mV	88%	–	Cao (2009)
						20		94%		
						35		93%		
Batch	Graphite brush 7 cm^2		Acetate 1 g L^{-1}	C. cloth 7 cm^2	Ferricyanide 50 mM	0.17 mol L^{-1} Ionic solution	Current <0.05 mA	1.07 mmol L^{-1} h^{-1}	–	Chen (2013)
								0.02 mmol L^{-1} h^{-1}		
Batch	C. cloth 7 cm^2		Dye	C. cloth 7 cm^2	Ferricyanide + PBS**	35	E_{cell} <30 mV	62%	35–90%	Kalleary (2014)
Batch, recirculation	C. brush		Acetate 4 g L^{-1} 0.17 mL min^{-1}	C. brush	Ferricyanide + PBS** 0.17 mL min^{-1}	5	–	100% conductivity	–	Zhang and He (2012)
						10		70%		
						20		41%		
Batch, recirculation (3 streams) 95 mL min^{-1}	100 cm^2	Isostatic graphite plate + C. felt	Sodium acetate	Isostatic graphite plate + C. felt	Ferricyanide	7.3	0.5 g L^{-1} NaCl	93.3%	7.14 kg COD m^{-3} day^{-1}	Ramírez-Moreno (2019)
						35	0.5 g L^{-1} NaCl	98.6%	19.7 kg COD m^{-3} day^{-1}	

	Stainless steel + C. fibre felt	Sodium acetate (2.5 $g\ L^{-1}$)	Stainless steel + C. fibre mesh + C. nanofibres doped iron nanoparticles	O_2/0.1 M PBS**	10.7	0.5 $g\ L^{-1}$ NaCl	93.6%	0.94 kg COD $m^{-3}\ day^{-1}$	
					33.5	17.4 $g\ L^{-1}$ NaCl	48.2%	1.07 kg COD $m^{-3}\ day^{-1}$	
4 MDCs series continuous flow	Treated C. brush 7 cm^2	Xylose 1 $g\ L^{-1}$ 0.25 mL min^{-1}	Pt/air-cat 7 cm^2	O_2/PBS**	20 $g\ L^{-1}$ NaCl (0.04 mL min^{-1})	5 $g\ L^{-1}$	76% 5.2 $g\ L^{-1}\ day^{-1}$	60% (1.15 kg COD $m^{-3}\ day^{-1}$)	Qu (2013)
					20 $g\ L^{-1}$ NaCl (0.02 mL min^{-1})	1 $g\ L^{-1}$	97% 3.36 $g\ L^{-1}\ day^{-1}$		
					20 $g\ L^{-1}$ NaCl (1 mL min^{-1})	–	–	–	
Batch	C. brush 9 cm^2	Acetate 3.0 $g\ L^{-1}$	Air cathode Fe-NCB*** 7 cm^2	O_2/PBS** (0.023 M)	30 $g\ L^{-1}$ NaCl	20 mS cm^{-1}	55% conductivity	83%	Santoro (2017)

*Calculated from published literature.

**PBS: Phosphate buffer solution.

***NCB: Nicarbazin (see more details in Santoro *et al.*, 2017).

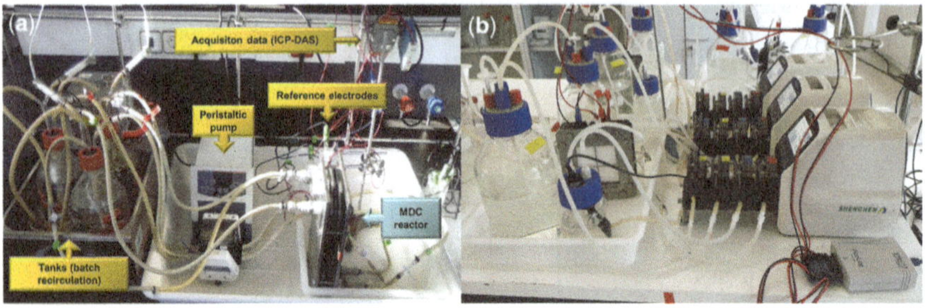

Figure 2.3 Pictures of the MDC system. Left: IMDEA Water. Right: LEITAT. Source: IMDEA Water Institute/LEITAT.

steel screws in order to avoid any leakage of the system. Figure 2.4 shows a flow diagram of the MDC experimental set-up.

Table 2.3 shows the constructive materials and the operational conditions used for the cathode strategies comparison.

2.4.1 MDC using liquid catholyte: experimental results at lab scale (IMDEA water facilities)

The MDC system was able to produce desalinated water with a simplified start-up procedure compared to other experimental systems reported in the literature, by adapting the microorganism before starting the MDC operation, thus significantly reducing the start-up period. Figure 2.5 shows the conductivity decrease for several experiments with various initial concentrations of NaCl. Table 2.4 summarizes the main results obtained.

Complete desalination was accomplished by the MDC lab set-up at IMDEA facilities. Note that in the literature it is reported that only partial desalination of seawater is possible using MDC technology. Secondly, it is important to note that production of desalinated water using liquid catholyte has reached a higher valued than reported in the literature (0.077 L m^{-2} h^{-1} for partial desalination from 52.4 to 20 mS cm^{-1}) (Zhang & He, 2015). Experimental results show that MDC technology is able to produce up 0.82 L m^{-2} h^{-1} for complete desalination of seawater.

2.4.2 MDC using air diffusion electrode as cathode (LEITAT facilities)

2.4.2.1 Air-cathode MDC test

MDC with air-diffusion cathode was tested under biotic conditions. For this experiment, carbon nanofibres doped with Fe nanoparticles produced by LEITAT

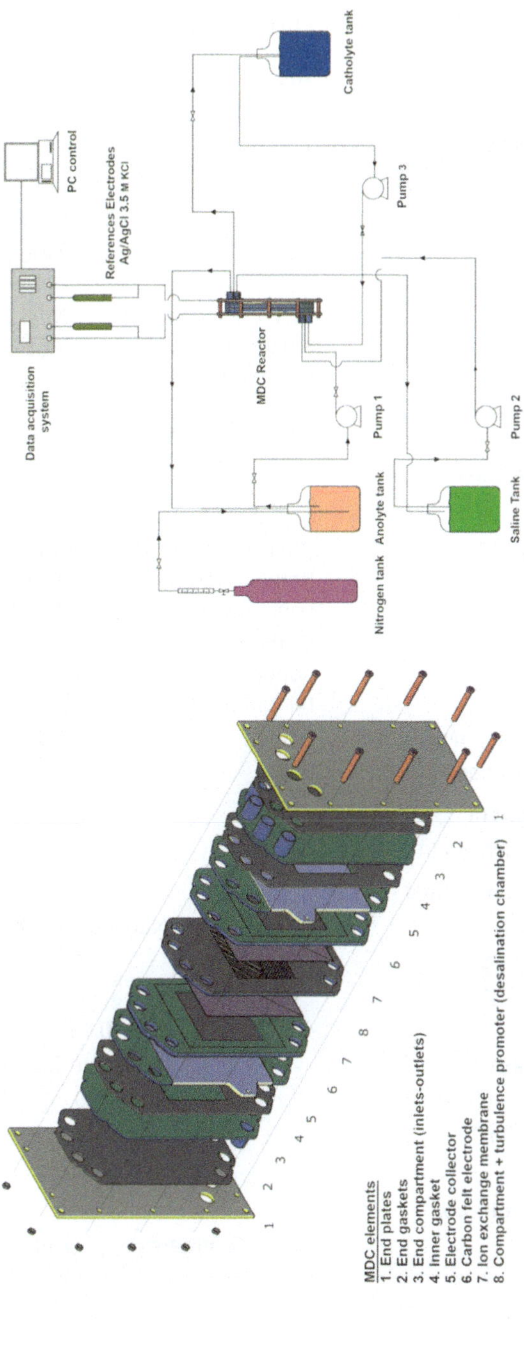

Figure 2.4 Diagram of an MDC unit (left) and flow diagram of MDC experimental set-up (right). Source: IMDEA Water Institute.

Table 2.3 Main features for MDC experimental setups.

Location	LEITAT Lab	IMDEA Water Lab
Cross section (cm^2)	100	100
Dimensions active area (cm)	10 × 10	10 × 10
Number of unit cells	1	1
Anode		
Electric collector	Stainless steel frame	Isostatic graphite plate (Grade 2114-45)
Electrode	SGL unidirectional carbon fibre felt/carbon felt	RVG 2000 MERSEN carbon felt
Electrode thickness (mm)	<0.5 (carbon fibres) – 2.5 (carbon felt)	4.6
Compartment thickness (mm)	8.7	9
Cathode		
Electric collector	Stainless steel 316 frame + carbon fibres mesh	Isostatic graphite plate (Grade 2114-45)
Electrode	Carbon nanofibres doped with iron nanoparticles	RVG 2000 MERSEN carbon felt
Electrode thickness (mm)	0.6	4.6
Compartment thickness (mm)	8.7	9
Saline compartment		
Compartment thickness (mm)	8.7	9
Ion exchange membranes		
Anionic membrane	Neosepta AMX	
Electric resistance (Ω cm^2)*	2.4	
Permselectivity (%)**	>93	
Thickness (μm)*	0.14	
Cationic membrane	Neosepta CMX	
Electric resistance (Ω cm^2)*	3.0	
Permselectivity (%)**	>90	
Thickness (μm)*	0.17	
Operational conditions		
Operation mode	Batch (3 streams)	Batch (3 streams)

(Continued)

Table 2.3 Main features for MDC experimental setups (*Continued*).

Location	LEITAT Lab	IMDEA Water Lab
Flow rate (mL min^{-1})	95	95
Temperature (°C)	25–30°C	25–30°C
External load (Ω)	2.5	2.5
Streams		
Anolyte	FWM*** + 2.5 g L^{-1} sodium acetate	FWM*** + sodium acetate
Catholyte	0.1 M PBS	$K_3[Fe(CN)_6]$
Saline stream	NaCl	NaCl
Tanks		
Anolyte volume (mL)	2500	2150
Catholyte volume (mL)	2500	2150
Saline volume (mL)	500	370
Rate V_{an}:V_{saline}:$V_{catholyte}$	5:1:5	5:1:5
Start-up operation		
Initial inoculum	Electroactive biofilm from an operating MFC	Pure culture *Geobacter sulfurreducens*
Period (hours)	158	140

*Equilibrated with 0.5N-NaCl solution, at 25°C (data provided by the manufacturer).
**Measured at the laboratory. Membrane equilibrated with 0.1 and 0.5 M NaCl solutions.
***FWM: Freshwater media (see more details in Borjas *et al.*, 2015).

(see Chapter 3) were used as the catalyst material in air-cathodes. Initial salt concentration in saline water was NaCl 10 g L^{-1} for brackish water and NaCl 35 g L^{-1} for seawater, but also real brackish water streams were tested. Salt removal achieved 99% and 48.3%, respectively, for 10 and 35 g NaCl L^{-1} experiments, as shown in Figure 2.6. No damage to ion exchange membranes (IEMs) was observed.

Table 2.5 summarizes the main results obtained. It can be observed that the nominal desalination rate (NDR) for brackish water is approximately 0.26 L m^{-2} h^{-1}, indicating that the air-diffusion cathode is a feasible strategy for complete brackish water desalination using MDC technology. For a real brackish water containing a mix of ions, with a conductivity comparable to 10 g L^{-1} of NaCl, the NDR decreased to 0.22 and only partial desalination was achieved (73%). On the other hand, a real brackish water with a conductivity comparable to 1.68 g L^{-1} of NaCl was successfully desalinated with an NDR

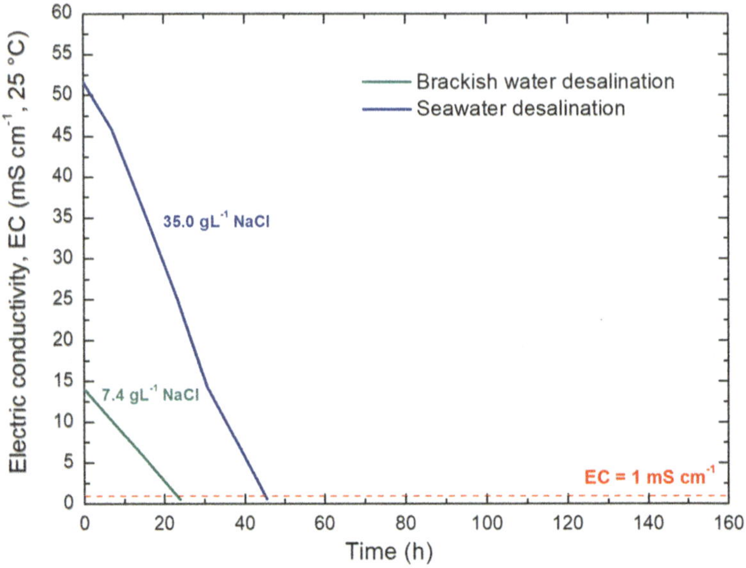

Figure 2.5 Electric conductivity in the saline tank for biotic desalination using liquid catholyte MDC. Horizontal line shows the limit for conductivity of 1 mS cm^{-1}. Source: IMDEA Water Institute.

of 0.66 L m^{-2} h^{-1}. In the case of seawater, partial desalination (48.2%) was obtained with this experimental set-up.

2.4.3 MDC using air diffusion cathode vs liquid catholyte: results comparison of brackish and seawater desalination

Table 2.6 shows the comparison of the results obtained using air diffusion cathode vs liquid catholyte, for comparison purposes of both MDC strategies (Ramírez-Moreno *et al.*, 2019).

2.5 DESIGN, CONSTRUCTION AND OPERATION OF MDC PRE-PILOT UNIT

Once the design of the MDC lab-scale set-up (cross section: 100 cm^2, 1 unit cell) had been validated, the system was scaled up to pre-pilot scale (cross section: 600 cm^2, 15–20 unit cells). The main purpose was to compare the performance of the pre-pilot MDC with the results obtained at lab scale, as well as to gain experience for further design of MDC pilot plants (operation at real environments).

Figure 2.7 shows the sequence of the assembly procedure for the manufacture of the MDC pre-pilot stack (end plates; gaskets; compartments; membranes; current

Table 2.4 MDC performance at different initials salinities for MDC lab set-up using liquid catholyte.

Salinity (g L^{-1}) NaCl	Saline Vol. (L)	Anolyte Vol. (L)	Catholyte Vol. (L)	Desalination Time (h) (Concentration <1 mS cm^{-1})	Nominal Desalination Rate (L m^{-2} h^{-1})	Current Efficiency η_C (%)	COD Removal Rate (kg COD m^{-3} day^{-1})
7.3 (brackish water)	0.37	2.15	2.15	23	1.5	81	7.14
35.0 (seawater)	0.37	2.15	2.15	43	0.7	108	19.7

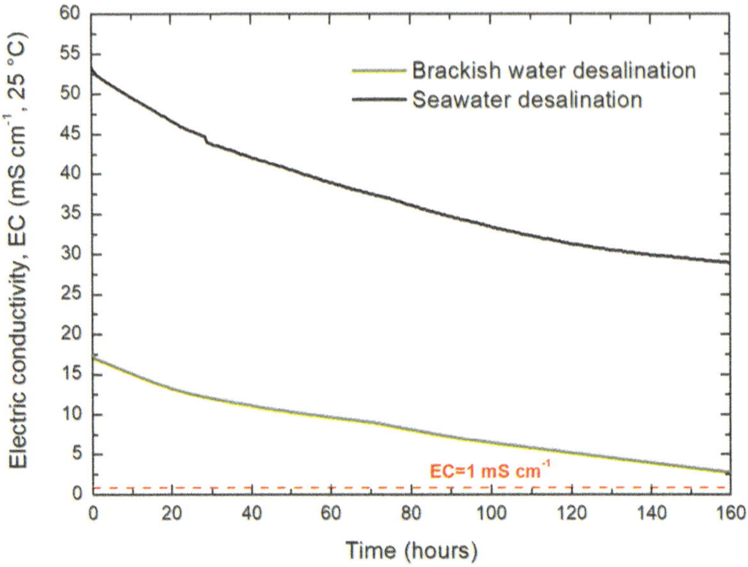

Figure 2.6 Electric conductivity in the saline tank for biotic desalination using in air-cathode MDC. Horizontal line shows the limit for conductivity of 1 mS cm^{-1}. Source: LEITAT.

collectors; electrodes; connection to auxiliary equipment: pumps, flow meters; integration of automatization and control system).

Figure 2.8 shows the final MDC pre-pilot setup. It is important to note that pre-pilot MDC operation is useful to support validation at pilot-plant scale, as it can be operated with similar experimental conditions to those on the real site. Indeed, the gained knowledge on the bioelectrochemical behaviour of the system (i.e., electric potential of the electrodes during desalination cycle, influence of the salinity on the process performance, analysis of the main parameters for proper operation of the system, performance of the anode biofilm in non-sterile conditions, etc.) was essential for the implementation of MDC technology in demo sites #1 and #2, as described in the next chapters.

The next paragraphs include the main results for brackish water and seawater,

2.5.1 Experiments with brackish water

Table 2.7 includes the main results for real brackish desalination using the pre-pilot MDC system. The use of synthetic and industrial wastewater as the source of COD has been tested to study the benefits of the technology as a desalination technology and water treatment technology. The experiments were carried out using ferro-ferricyanide catholyte (liquid catholyte).

Table 2.5 MDC performance at different initials salinities for MDC lab set-up using air-diffusion cathode.

Salinity (g L^{-1}) NaCl	Saline Vol. (L)	Anolyte Vol. (L)	Catholyte Vol. (L)	Desalination Time (h) (Concentration <1 mS cm^{-1})	Nominal Desalination rate (L m^{-2} h^{-1})	Current Efficiency η_C (%)	COD Removal Rate (g COD m^{-3} day^{-1})
10 (synthetic brackish water)	0.5	2.5	2.5	199	0.26	162	467.4
10 (brackish water)	0.5	2.5	2.5	168	0.22 (for partial desalination 73%)	104	307.5
35.0 (synthetic seawater)	0.5	2.5	2.5	168	0.14 (for partial desalination 48.2%)	145	536.7
1.68 (real brackish water)	0.5	2.5 (industrial wastewater)	2.5	75	0.66	39.1	346.56

Table 2.6 Initial and final salinity, salt removal, desalination time, current efficiency, nominal desalination rate, COD removal rate, anode coulombic efficiency, specific energy production, total circulated charge and volume variation for air diffusion cathode and liquid catholyte experiments for brackish water and seawater desalination.

	Air-Diffusion Cathode MDC		Liquid Catholyte MDC	
	Brackish Water	**Seawater******	**Brackish Water**	**Seawater**
Initial salinity (g L^{-1})	10.7	33.5	7.3	35.0
Final salinity (g L^{-1})	0.5	17.4	0.5	0.5
Salt removal (%)	93.6	48.2	93.3	98.6
Desalination time (h)	205	–	23	43
NDR (L m^{-2} h^{-1})*	0.26**	0.14	1.5	0.7
Current efficiency (%)	162	145	81.1	108
COD removal rate (kg COD m^{-3} day^{-1})***	0.47	0.54	7.14	19.7
Coulombic efficiency (%)	6.5	10.3	91.0	61.0
Specific energy production (kWh m^{-3})	0.017	0.055	0.7	5.4
Total circulated charge (Q)	5086	8994	5165	19354
Volume variation (%)	−36.0	−2	−8.1	−19

*Calculated considering the final volume of saline tank.

**Extrapolated from experimental results.

***Considering the volume of anolyte compartment.

****Calculated for partial desalination.

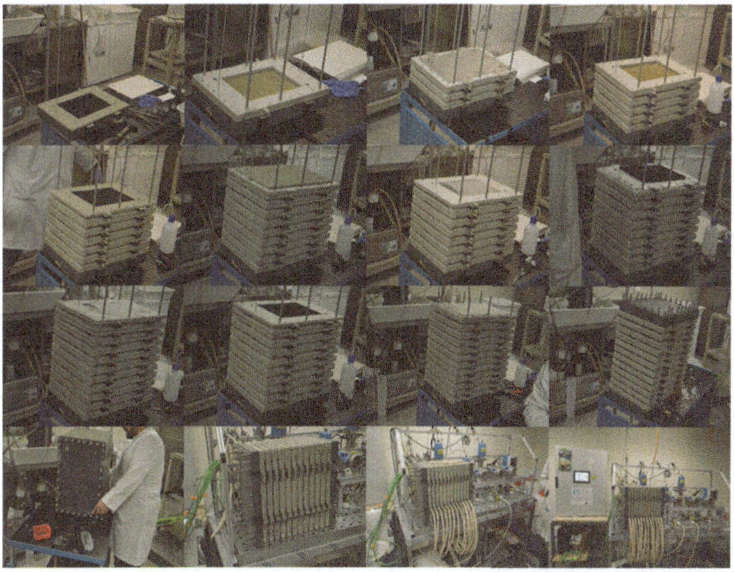

Figure 2.7 Sequence of the MDC stack assembly. Source: IMDEA Water Institute.

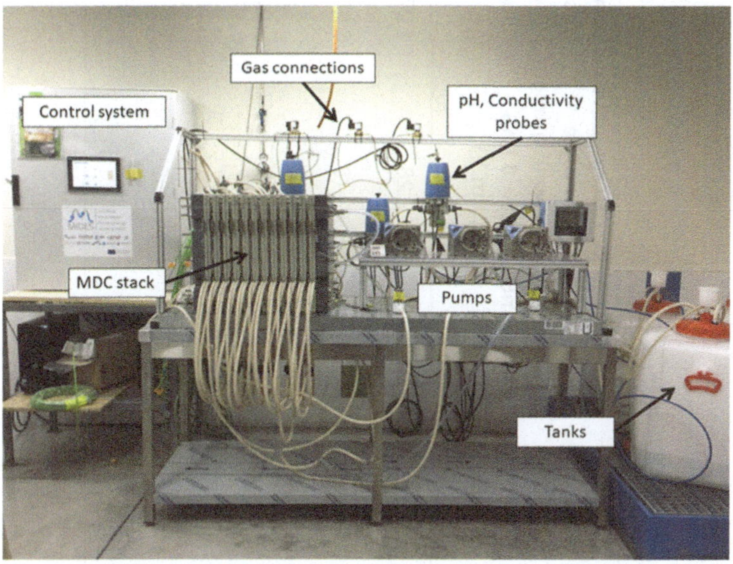

Figure 2.8 Pre-pilot MDC system at IMDEA facilities. MDC pre-pilot system: 600 $cm^2 \times$ 10–20 units (6000–12,000 cm^2, modular system). Membranes: Neosepta AMX/CMX. Electrodes: Isostatic graphite/Carbon felt: KF4.5 (both provided by SGL Carbon). WW: Brewery wastewater (Cond = 11 mS cm^{-1}), Catholyte: $K_3Fe(CN)_6$ (Cond = 20 mS cm^{-1}). Source: IMDEA Water Institute.

Table 2.7 Main results for MDC pre-pilot system for brackish water desalination.

Id.	Saline Solution	Anolyte Solution	Conductivity Saline Compartment t ≈ 0 (mS cm^{-1})	Conductivity Anolyte Compartment t ≈ 0 (mS cm^{-1})	Nominal Desalination Rate (NDR) (L m^{-2} h^{-1})	Current Efficiency %	Coulombic Efficiency %	Specific Electric Production kWh m^{-3}	COD Removal Rate kg COD m^{-3} day^{-1}
1	NaCl 5 g L^{-1}	Synthetic FWM	10.60	9.28	1.27	44.8	32.41	1.706	8.45
2	NaCl 1.75 g L^{-1}	Synthetic FWM	3.16	9.16	6.25	31.8	31.27	0.691	13.70
3	Racons river water	Synthetic FWM	2.20	9.16	11.24	33.1	57.39	0.409	7.26
4	Racons river water	Industrialwater (food industry)	2.92	2.55	3.9	40.3	52.51	0.243	3.56

Table 2.8 Main results for MDC pre-pilot system for seawater desalination.

Id.	Saline Solution	Anolyte Solution	Conductivity Saline Compartment $t \approx 0$ (mS cm^{-1})	Conductivity Anolyte Compartment $t \approx 0$ (mS cm^{-1})	Nominal Desalination Rate (NDR) (L m^{-2} h^{-1})	Current Efficiency %	Coulombic Efficiency %	Specific Electric Production $kWhm^{-3}$	COD Removal Rate kg COD m^{-3} day^{-1}
5	NaCl 35 g L^{-1}	Synthetic FWM	46.4	8.36	0.42	152.0	27.61	0.202	7.46
6	Seawater	Synthetic FWM	28.32	8.73	0.59	90.7	21.58	0.691	15.36
7	Seawater	Industrial water (food industry)	2.20	9.16	0.14	33.1	57.39	0.409	7.26

It is important to indicate that NDR for brackish water desalination is 3.9 L $m^{-2} h^{-1}$ for real brackish and real wastewater (food industry, brewery). The difference between this value and 11.24 L $m^{-2} h^{-1}$ (using synthetic wastewater as anolyte) is mainly due to the difference in the initial conductivity of the anolyte (synthetic: 9.16 mS cm^{-1} and wastewater: 2.55 mS cm^{-1}), indicating that this parameter plays a key role in MDC performance. The lower electric conductivity in the real wastewater impacts on the electrochemical performance of the MDC system, increasing the internal resistance and reducing the energy available to drive the desalination process. This consideration should be taken into account for the validation of the technology when real wastewater is used as the feed for the MDC system. In general, the electric conductivity of the wastewater will influence the desalination process, with wastewater with electric conductivity in the range 6–9 mS cm^{-1} being optimal for the microbial desalination process. In any case, the MDC performance at pre-pilot scale represents a breakthrough for a technology up-scaled from laboratory to pre-pilot scale.

2.5.2 Experiments with seawater

The experiments performed with seawater are summarized in Table 2.8. Again, the experiments were carried out using ferro-ferricyanide catholyte (liquid catholyte).

The experimental results for seawater desalination show nominal desalination rates (NDR) in the range of 0.1 to 0.6 L $m^{-2} h^{-1}$, these values are comparable to the laboratory-scale system, showing good accordance for the up-scaled MDC system.

Figure 2.9 LEITAT air cathode implemented in the cathode compartment of the pre-pilot MDC system at IMDEA facilities (see details in Chapter 3). Source: IMDEA Water Institute/LEITAT.

Table 2.9 Main results for MDC pre-pilot system for brackish water and seawater desalination, using air diffusion cathode developed by LEITAT.

Id.	Saline Solution	Anolyte Solution	Conductivity Saline Compartment $t \approx 0$ (mS cm^{-1})	Conductivity Anolyte Compartment $t \approx 0$ (mS cm^{-1})	Nominal Desalination Rate (NDR) (L m^{-2} h^{-1})	Current Efficiency %	Coulombic Efficiency %	Specific Electric Production kWh m^{-3}	COD Removal Rate kg COD m^{-3} day^{-1}
8	NaCl 5 g L^{-1}	Synthetic FWM	9.26	9.35	0.16	54.0	2.31	1.706	–
9	NaCl 2.5 g L^{-1}	Synthetic FWM	4.85	9.52	0.31	61.2	1.48	0.691	–
10	Racons river water	Synthetic FWM	6.08	7.41	0.30	61.1	1.42	0.409	26.24
11	Seawater	Synthetic FWM	32.5	7.81	0.05	62.7	1.95	0.243	2.15

2.5.3 Experiments in pre-pilot MDC system using air diffusion cathode (collaboration LEITAT-IMDEA)

An air cathode is a promising development for sustainable desalination using microbial desalination cells (see Chapter 3). In this sense, LEITAT developed a large-scale air diffusion cathode for implementation in the pre-pilot MDC system (unit cell) based on their lab-scale experience (Figure 2.9).

Experiments were performed to compare the air cathode experiments with the experiments already performed with the liquid catholyte (see previous sections). Table 2.9 summarizes the obtained results.

The obtained results with the air diffusion cathode designed by LEITAT demonstrate that is possible to produce fresh water from brackish and seawater. The nominal desalination rates (NDR) are slightly lower compared to the liquid catholyte, but results show promising development for future implementation of this renewable cathode on future systems.

2.6 SUMMARY

The main goal of this chapter is to present the main concepts and principles for the microbial desalination process developed in MIDES Project. Main results at lab and pre-pilot scale are presented in the chapter. More information related to the electrodes and membranes used in this chapter can be found in Chapter 3. The obtained results were the starting point for the design and construction of the MDC pilot plants, as indicated in Chapter 4, and operation and validation of MIDES technology (see Chapter 6). Also, they were partially used for validation of the mathematical modelling simulation (see Chapter 5) and LCA study (Chapter 7).

REFERENCES

Borjas Z., Ortiz J. M., Aldaz A., Feliu J. and Esteve-Núñez A. (2015). Strategies for reducing the start-up operation of microbial electrochemical treatments of urban wastewater. *Energies*, **8**, doi: 10.3390/en81212416

Cao X., Huang X., Liang P., Xiao K., Zhou Y., Zhang X. and Logan B. E. (2009). A new method for water desalination using microbial desalination cells. *Environmental Science Technology*, **43**, 7148–7152, doi: 10.1021/es901950j

Chen X., Xia X., Liang P., Cao X., Sun H. and Huang X. (2011). Stacked microbial desalination cells to enhance water desalination efficiency. *Environmental Science Technology*, **45**, 2465–2470, doi: 10.1021/es103406 m

Chen X., Liang P., Wei Z., Zhang X. and Huang X. (2012). Sustainable water desalination and electricity generation in a separator coupled stacked microbial desalination cell with buffer free electrolyte circulation. *Bioresource Technology*, **119**, 88–93, doi: 10.1016/J.BIORTECH.2012.05.135

Chen S., Luo H., Hou Y., Liu G., Zhang R. and Qin B. (2013). Comparison of the removal of monovalent and divalent cations in the microbial desalination cell.

Frontiers Environmental Science Engineering, **9**, 317–323, doi: 10.1007/s11783-013-0596-y

ElMekawy A., Hegab H. M. and Pant D. (2014). The near-future integration of microbial desalination cells with reverse osmosis technology. *Energy Environmental Science*, **7**, 3921–3933, doi: 10.1039/C4EE02208D

Ge Z., Dosoretz C. G. and He Z. (2014). Effects of number of cell pairs on the performance of microbial desalination cells. *Desalination*, **341**, 101–106, doi: 10.1016/J.DESAL.2014.02.029

Jacobson K. S., Drew D. M. and He Z. (2011a). Efficient salt removal in a continuously operated upflow microbial desalination cell with an air cathode. *Bioresource Technology*, **102**, 376–380, doi: 10.1016/j.biortech.2010.06.030

Jacobson K. S., Drew D. M. and He Z. (2011b). Use of a liter-scale microbial desalination cell as a platform to study bioelectrochemical desalination with salt solution or artificial seawater. *Environmental Science Technology*, **45**, 4652–4657, doi: 10.1021/es200127p

Kalleary S., Mohammed Abbas F. M., Ganesan A., Meenatchisundaram S., Srinivasan B., Packirisamy A. S. B., Kesavan R. K. and Muthusamy S. (2014). Biodegradation and bioelectricity generation by microbial desalination cell. *International Biodeterioration Biodegradation*, **92**, 20–25, doi: 10.1016/J.IBIOD.2014.04.002

Kim Y. and Logan B. E. (2011a). Series assembly of microbial desalination cells containing stacked electrodialysis cells for partial or complete seawater desalination. *Environmental Science Technology*, **45**, 5840–5845, doi: 10.1021/es200584q

Logan B. E., Wallack M. J., Kim K.-Y., He W., Feng Y. and Saikaly P. E. (2015). Assessment of microbial fuel cell configurations and power densities. *Environmental Science Technology Letters*, **2**, 206–214, doi: 10.1021/acs.estlett.5b00180

Mehanna M., Saito T., Yan J., Hickner M., Cao X., Huang X. and Logan B. E. (2010). Using microbial desalination cells to reduce water salinity prior to reverse osmosis. *Energy Environmental Science*, **3**, 1114–1120, https://doi.org/10.1039/c002307h

Ping Q., Cohen B., Dosoretz C. and He Z. (2013). Long-term investigation of fouling of cation and anion exchange membranes in microbial desalination cells. *Desalination*, **325**, 48–55, doi: 10.1016/J.DESAL.2013.06.025

Qu Y., Feng Y., Wang X., Liu J., Lv J., He W. and Logan B. E. (2012). Simultaneous water desalination and electricity generation in a microbial desalination cell with electrolyte recirculation for pH control. *Bioresource Technology*, **106**, 89–94, doi: 10.1016/j.biortech.2011.11.045

Qu Y., Feng Y., Liu J., He W., Shi X., Yang Q., *et al.* (2013). Salt removal using multiple microbial desalination cells under continuous flow conditions. *Desalination*, **317**, 17–22, doi: 10.1016/j.desal.2013.02.016

Ramírez-Moreno M., Rodenas P., Aliaguilla M., Bosch-Jimenez P., Borràs E., Zamora P., Monsalvo V., Rogalla F., Ortiz J. M., Esteve-Núñez A. (2019). Comparative performance of microbial desalination cells using air diffusion and liquid cathode reactions: study of the salt removal and desalination efficiency. *Frontiers Energy Research*, **7**, 135, doi: 10.3389/fenrg.2019.00135

Saeed H. M., Husseini G. A., Yousef S., Saif J., Al-Asheh S., Abu Fara A., Azzam S., Khawaga R. and Aidan A. (2015). Microbial desalination cell technology: a review and a case study. *Desalination*, **359**, 1–13, doi: 10.1016/J.DESAL.2014.12.024

Santoro C., Talarposhti M. R., Kodali M., Gokhale R., Serov A., Merino-Jimenez I., Ieropoulos I. and Atanassov P. (2017). Microbial desalination cells with efficient platinum-group-metal-free cathode catalysts. *ChemElectroChem*, **4**, 3322–3330, doi: 10.1002/celc.201700626

Schröder U., Harnisch F. and Angenent L. T. (2015). Microbial electrochemistry and technology: terminology and classification. *Energy Environmental Science*, **8**, 513–519, doi: 10.1039/C4EE03359 K

Wen Q., Zhang H., Chen Z., Li Y., Nan J. and Feng Y. (2012). Using bacterial catalyst in the cathode of microbial desalination cell to improve wastewater treatment and desalination. *Bioresource Technology*, **125**, 108–113, doi: 10.1016/J.BIORTECH.2012.08.140

Yuan H., Abu-Reesh I. M. and He Z. (2015). Enhancing desalination and wastewater treatment by coupling microbial desalination cells with forward osmosis. *Chemical Engineering Journal*, **270**, 437–443, doi: 10.1016/j.cej.2015.02.059

Zhang B. and He Z. (2012). Integrated salinity reduction and water recovery in an osmotic microbial desalination cell. *RSC Advances*, **2**, 3265, doi: 10.1039/c2ra20193c

Zhang F. and He Z. (2015). Scaling up microbial desalination cell system with a post-aerobic process for simultaneous wastewater treatment and seawater desalination. *Desalination*, **360**, 28–34, doi: 10.1016/j.desal.2015.01.009

Zhang F., Chen M., Zhang Y. and Zeng R. J. (2012). Microbial desalination cells with ion exchange resin packed to enhance desalination at low salt concentration. *Journal Membrane Science*, **417–418**, 28–33, doi: 10.1016/J.MEMSCI.2012.06.009

Zuo K., Chang J., Liu F., Zhang X., Liang P. and Huang X. (2017). Enhanced organics removal and partial desalination of high strength industrial wastewater with a multi-stage microbial desalination cell. *Desalination*, **423**, 104–110, doi: 10.1016/J.DESAL.2017.09.018

Zuo K., Chen M., Liu F., Xiao K., Zuo J., Cao X., *et al.* (2018). Coupling microfiltration membrane with biocathode microbial desalination cell enhances advanced purification and long-term stability for treatment of domestic wastewater. *Journal Membrane Science*, **547**, 34–42, doi: 10.1016/J.MEMSCI.2017.10.034

Chapter 3

Key elements and materials in microbial desalination cells

Eduard Borràs[1], *Martí Aliaguilla*[1], *Laura Huidobro*[1], *Sandra Martínez-Crespiera*[1], *Sonia Matencio*[1], *Daniele Molognoni*[1], *Juan Manuel Ortiz*[2], *Marina Ramírez-Moreno*[2], *Pau Rodenas*[2], *Mirko Faccini*[1], *Marta Juan-y-Seva*[1], *Ruediger Schweiss*[3], *Almut Schwenke*[3], *Maarten Meijlink*[4], *Abdulsalam Alhadidi*[4] and *Pau Bosch-Jimenez*[1]

[1]*LEITAT Technological Center, C/ de la Innovació, 2, 08225 Terrassa, Barcelona, Spain*
[2]*IMDEA Water, Av/ Punto Com, 2, Parque Científico Tecnológico de la Universidad de Alcalá, 28805, Alcalá de Henares, Madrid, Spain*
[3]*SGL Carbon GmbH, Werner-von-Siemensstrasse 18, 86405 Meitingen, Germany*
[4]*FUJIFILM Manufacturing Europe B.V. P.O. Box 90156, 5000 LJ Tilburg, The Netherlands*

ABSTRACT

This chapter presents the most relevant advances achieved during the MIDES project in relation to material development of key elements for microbial desalination cells. The first section is devoted to electrodes. Providing a general overview of the requirements of carbon-based materials to serve

doi: 10.2166/9781789062120_0041

either as anodes or cathodes for microbial desalination cells. Advances achieved during MIDES in the development of materials for anode and cathode application are listed. The second section is focussed on ion-exchange membranes for microbial desalination cells. General considerations for the use of these membranes are reported as well as key parameters. Finally, advances in ion-exchange membrane development, in terms of antifouling and their performance in desalination trials, achieved during the MIDES project, are reported.

Keywords: Air-cathode, anion-exchange membrane, anode, carbon materials, cation-exchange membrane, electrodes, liquid cathode.

3.1 ELECTRODES

3.1.1 General considerations

Electrodes constitute essential components in microbial desalination cells (MDCs) as they determine the magnitude of the potential gradient which drives the desalination process. This chapter summarizes the development of low-cost, scalable, carbon-based collector and electrode materials for the microbially-assisted desalination cells used in MIDES. **Collector materials** serve as the structural unit of the MDCs which facilitate electrical wiring of electrodes and control electronics. **Anode materials** are porous, conductive 3D structures which host the electroactive biofilms which metabolise organic matter. **Cathode materials** perform electrochemical reduction of oxidant species (oxygen or dissolved redox active species) thereby producing positive potentials which drive the desalination process. Graphite and carbon materials are the perfect choice for the aforementioned uses due to the following properties (Yuan *et al.*, 2016; Zhou *et al.*, 2011):

- Biocompatibility
- Corrosion resistance
- Availability with high purity
- Availability in a variety of shapes
- Processability as textile (roll-to-roll) or polymer compound materials (hot-pressing, extrusion)
- Moderate manufacturing cost and industrial scalability

3.1.1.1 Electronic conductivity

Despite the rather low currents flowing in MDCs, a sufficient electronic conductivity of the porous materials is required to obtain a high utilization of the electrode. Carbon fibre conductivity can be tuned by means of the nature of the carbon fibre precursor and by the temperature applied during inert-gas pyrolysis of the precursors to carbon. For the current collectors, stainless-steel plates or thin composite sheets based on expanded graphite/fluorinated polymers are available as low cost materials. Typically, the main voltage losses in such a cell design are related to the contact resistance of the carbon fibre electrode and the graphite plate. In practice, this fact is mitigated by a certain compression of the fibre-based electrode (around 20–30%) by means of the flow frame since the fibre matrix shows an elastic behaviour under such conditions ('spring property').

3.1.1.2 Surface area

Commercially available carbon-fibre electrodes such as felt or cloths show a favourably large surface area measured by the Brunauer, Emmett and Teller technique (BET, of 0.3 to 1 m^2/g equivalent to a volumetric surface area of $>250\ cm^3/cm^2$) which allows for a high conversion level in flow-through reactors. This property is also important with respect to MDCs, and to bioelectrochemical systems reactors in general terms, in order to achieve efficient wiring of the bacteria onto the electrode surface. Surface treatments like thermal oxidation or plasma etching have already been shown to enhance the surface area by a factor of 3 to 10 (Ghimire *et al.*, 2019).

3.1.1.3 Surface chemistry

The carbon fibre precursor and the carbonization procedure have effects on the surface area and the surface chemistry of the fibre. Moderate temperatures during fabrication help to retain a larger number of functional groups such as nitrogen species (PAN [polyacrylonitrile] precursor) and surface oxygen complexes (PAN and rayon precursor).

Microbial adhesion to a surface, which is highly dependent on the physicochemical properties of both the cells and surfaces, is an essential step in electroactive biofilm formation. Surface modification therefore represents an important way to modulate microbial attachment and ultimately biofilm formation by electroactive microorganisms. Different surface modification processes, such as organosilane surface modification, plasma treatment, chemical modification of carbon nanotubes, electro-oxidation and covalent-immobilization with neutral red and methylene blue molecules, have been previously studied in literature (Mano, 2020). The effectiveness of these surface modifications, as a process to enhance beneficial biofilm formation, and their industrial application have been scarcely studied so far. These methods need to be safe, economically viable, scalable and environmentally friendly, and their potential to fulfil these criteria for many applications is yet to be determined.

3.1.1.4 Anode specifications

Since the redox reaction on the anode of the MDC is governed by the ability of the electroactive biofilm to digest organic matter and to deliver electrons to the carbon electrode, the main focus is to provide sufficient surface area and to provide a functional biofilm on the carbon matrix. Commercial carbon felt with porosity larger than 90% should constitute a suitable platform technology for the anode of the MDC. Moreover, special attention will be paid to the influence of differing surface chemistry in order to determine optimum performance of the bioanode in the MDC system. A better understanding of how anode surface properties affect growth, development, and activity of electrogenic biofilms has great potential to improve the performance of the MDC. Carbon-based materials are widely used for electrode materials and can vary in surface properties such as morphology, chemistry, available surface area and wettability. Attachment surface charge and wettability (hydrophilicity/hydrophobicity) may be particularly important to the growth, development and properties of bacterial biofilm and optimization of MDC performance.

3.1.1.5 Cathode specifications

The analysis of the literature shows that two strategies (liquid catholyte and air-cathode) are the most promising options for an optimal configuration of the MDC (Ramírez-Moreno *et al.*, 2019).

A complementary cathode reaction is required to sustain the electrochemical reaction of the cell. If the corresponding redox system is also a liquid-liquid couple (Fe^{2+}/Fe^{3+}or ferro/ferricyanide) in the so-called symmetric MDC (Cao *et al.*, 2009), a carbon felt with geometric and pore dimensions similar to the anode could be employed. Matching of the surface chemistry of the felt to the corresponding redox species could be used to optimize the electrochemical kinetics of the cathode.

When using the oxygen reduction reaction (ORR) on the cathode side, called air-cathode, the preferred cathode design is similar to a gas diffusion electrode coated with (noble) metal catalyst (Bhattacharjee & Pandey, 2020; Majidi *et al.*, 2019). Thin carbon paper-based electrodes (0.25 mm) used in electrochemical power sources (proton exchange membrane (PEM) fuel cells, metal-air batteries) could be adapted for that purpose. The thin carbon papers provide a short diffusion path for the oxygen and a suitable hydrophobic treatment provides a 3-phase boundary for the electrochemical reactions. Additionally, such fibre materials could be easily coated with carbon-supported catalytically active materials (preferably catalysts which are not based on precious metals). Alternatively, catalytically active materials could be deposited directly onto the fibre materials. This concept is attractive as the use of air as oxidant at the positive electrode reduces the footprint of the MDC system. The critical issue, however, is related to the substantial overpotentials of the air-breathing cathodes.

3.1.2 Electrodes manufacturing

3.1.2.1 Anode materials

Carbon materials (particularly fibre-based) constitute excellent electrode materials for microbial electrochemical cells due to their biocompatibility and excellent

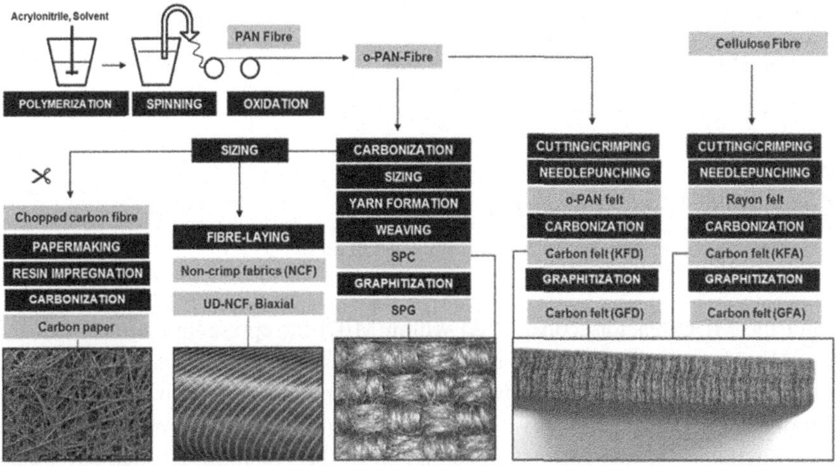

Figure 3.1 Manufacturing routes of different carbon fibre electrode structures suitable for MDC application. Source: SGL.

electronic conductivity. These come in a variety of forms/shapes and thus allow for a variety of different cell designs (Cai *et al.*, 2020). Within MIDES, emphasis was put on carbon fibre-based materials due to their high technical and commercial relevance.

Figure 3.1 presents a survey of the different production routes of carbon fibre structures which could be used potentially as electrode materials in MDC application. Carbon papers are thin flat structures (0.15 to 0.3 mm) which are formed from chopped carbon fibres which are bonded with carbonized resins. These are typical substrates for PEM fuel cell electrodes. Non-crimp fabrics are common in lightweight construction and are made by laying fibres (from spools) in a single direction (unidirectional – UD) or biaxially and by fixing with a polyester yarn. Woven fabrics are produced by weaving of carbon fibre yarns and are highly isotropic. Carbon felts are thick (1.5 to 10 mm), elastic materials, which are prepared by needle-punching carbon precursor fibres and subsequent carbonization/graphitization. They are very common in flow-through electrochemical reactors such as vanadium redox flow batteries (Huong Le *et al.*, 2017; Minke *et al.*, 2017). All types of carbon fibre materials are manufactured using high volume roll-to-roll (R2R) processes and enable materials to be made with different functional characteristics with respect to stiffness, smoothness, isotropy, flatness, compressional elasticity or porosity. The predominant precursor of carbon fibres (and related fabrics) is polyacrylonitrile (PAN) as it allows for the production of carbon fibres with higher crystallinity and conductivity.

Table 3.1 shows the area-specific resistances obtained for the different substrates:

Table 3.1 Typical area-specific resistance of selected carbon fibre materials.

Carbon Material	**Compressed Thickness (mm)**	**ASR (mΩ cm^2)**
Paper (36 AA)	0.22	4
Woven (SPG)	0.28	7
Felt (KFD2.5)	2.08	62

ASR: area-specific resistance at a compression ratio of 20%.

For the MDC reactors constructed within MIDES, representative materials for each manufacturing route have been tested for their application as anodes. A carbonized felt material (KFD) was preferred for the MDC as it allowed better performance of the electroactive biofilm. The amorphous fibre properties retain more functional surface groups (as compared to the graphitized materials) which promote electroactive biofilm formation.

3.1.2.2 Cathode materials

3.1.2.2.1 Liquid cathodes

Cathodes are based on a redox system dissolved in water. Hence, a 3D flow-through electrode is most effective. Due to the high surface area and low flow resistance, carbon felts (which have been used in electrochemical flow reactors for more than three decades) are most suitable. Redox couples such as iron (III) show good kinetics at carbon felt electrodes as demonstrated previously (Lopez-Atalaya *et al.*, 1992; Oh *et al.*, 2004). In Chapter 2, the use of liquid catholyte for MDC operation is described, as well as the main advantages and limitations compared to using an oxygen reduction reaction as the cathodic reaction.

3.1.2.2.2 Air-cathodes

Air-cathodes utilize the oxygen reduction reaction (ORR) to operate as positive electrodes in MDCs.

$$O_2 + 4H^+ + 4e^- \rightarrow 2H_2O (0.605\ \text{V vs. Ag/AgCl at pH 7})$$

Since the reaction has poor kinetics (particularly at neutral pH), effective electrocatalysts (high mass activity and surface area) are needed (Yuan *et al.*, 2016). Common catalysts used in proton exchange membrane fuel cells (PEMFCs) and alkaline fuel cells (AFCs) such as carbon supported platinum (PEMFC/acidic conditions) and silver or manganese dioxide (alkaline/AFC) could be adopted for the MDC. Similarly, novel catalysts based on heteroatom (Fe, Ni, Co)-doped carbons, which are potentially low-cost, were suggested as promising candidates for microbial cells (Merino-Jimenez, 2016). Moreover, such

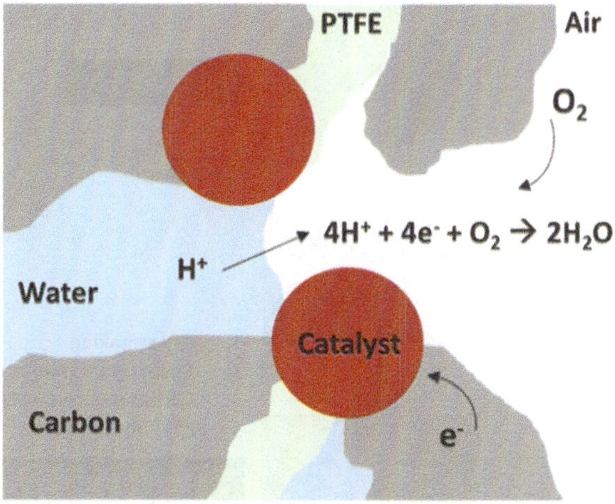

Figure 3.2 Three-phase boundary at the active sites of an air-cathode performing the oxygen reduction reaction (ORR). Source: Leitat and SGL.

an electrode requires a structure which maximizes the three-phase boundaries at the reactive sites to maintain the electrode reactions/potential. The catalyst particles must be accessible to both oxygen and liquid water as shown in Figure 3.2. On the other hand, the electrode should be waterproof, in order to prevent water leakage. This is typically accomplished by either (a) a PTFE fabric which is loaded with a carbon-supported catalyst or (b) a two-layer structure with a PTFE-impregnated carbon support, which is subsequently coated with a catalyst ink based on PTFE, carbon and catalyst particles.

3.1.2.2.2.1 Air cathodes via GDE approach. Gas diffusion electrodes (GDEs) have been used in fuel cells for several decades. Consequently, Pt/C gas diffusion electrodes, which are used in proton exchange membrane fuel cells (PEMFCs) and phosphoric acid fuel cells (PAFCs) have been employed as air-cathodes in MDCs (Santoro *et al.*, 2017). Two designs have been investigated within MIDES.

Type A uses a gas diffusion layer (GDL) which is made by coating a PTFE-impregnated carbon paper (5 wt% PTFE) with a carbon/PTFE microporous layer. On this GDL, a thin film consisting of carbon black, catalyst particles and perfluorosulfonic acid (Nafion®) ionomer (proton conducting polymer with low equivalent weight) is coated and annealed (120°C). This structure is very similar to GDEs used in PEMFCs.

Type B is made by direct coating an ink consisting of high surface area carbon black, catalyst particles and PTFE onto a carbon paper support similar to gas diffusion electrodes designed for AFCs (Figure 3.3).

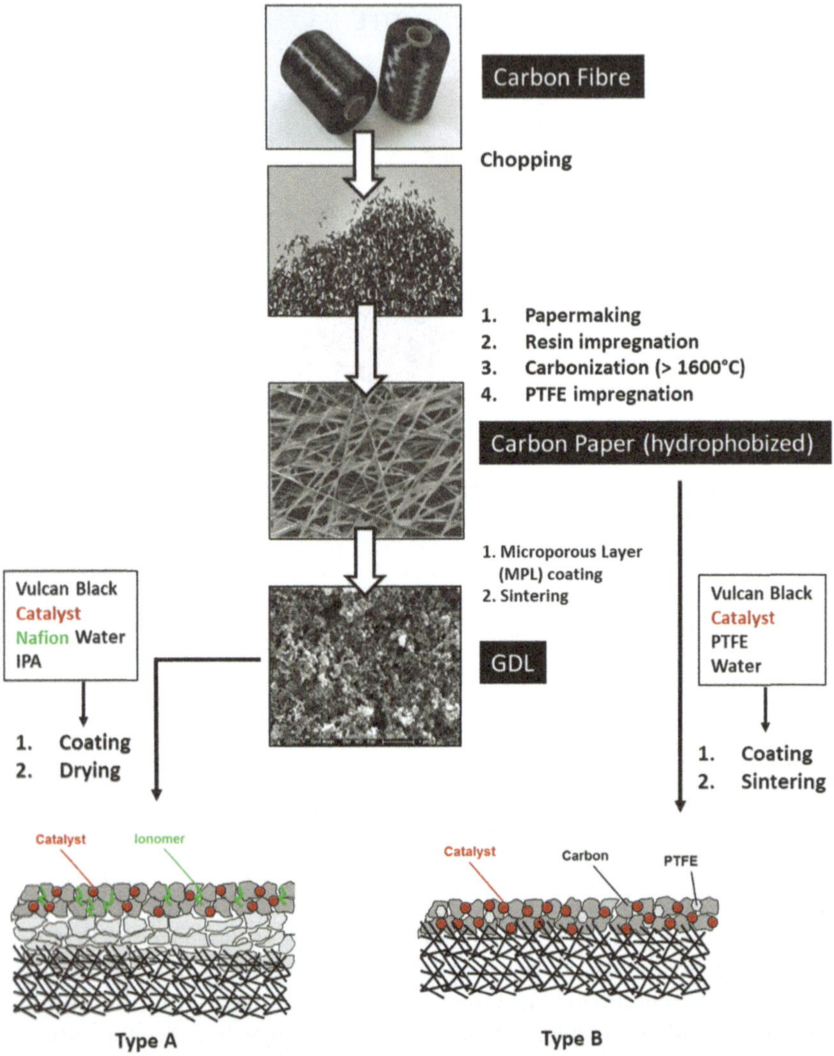

Figure 3.3 Manufacturing routes of GDE air-cathodes based on carbon paper supports. Source: SGL.

GDEs with various catalysts were investigated within MIDES:

- Pt on carbon cloth as reference
- Commercial non-platinum-group-material catalyst (NPGM)
- Commercial battery-grade, high surface area electrolytic manganese dioxide (EMD)

- SGL-made catalyst (iron-doped carbon made by pyrolysis of an iron-tannin complex mixed with Vulcan Black).

3.1.2.2.2.2 Air-cathodes based on electrospun carbon nanofibres. Carbon nanofibres (CNFs) exhibit very unique properties that make them promising candidates for catalysts in fuel cell applications, like MDCs. Ghasemi and coauthors (2011) demonstrated that the power per cost of activated CNFs fabricated by electrospinning and used in cathodes in MFCs was 2.65 times greater than that of the traditionally used platinum cathode. In another study, Santoro and coauthors (2013) demonstrated that the CNFs as cathodes in MFCs exhibited less deterioration throughout the operational period in comparison with Pt-based cathodes, which demonstrated their great potential as cost-effective cathodes for long-term operation.

The production process of CNFs involves three steps (Figure 3.4): electrospinning of polymer (or mix of polymers) solution doped with metal precursor (iron/cobalt acetate); stabilization at 280–300°C in air; and finally, carbonization under Ar atmosphere at 800–1000°C. After these treatments CNFs doped with nanoparticles (Co and Fe) generated in situ are obtained.

The first step to produce CNFs is the optimization of the electrospinning process of the precursor polymer (Figure 3.5). Variables related to the precursor solution (concentration, molecular weight (MW), viscosity, surface tension, solvent), the equipment parameters (voltage, distance sample-collector, flow rate, collector type and speed, needle) and the ambient conditions (temperature, relative humidity, air flow) influence the final electrospun nanofibre properties. First carbon nanofibres have been produced from 10–12 wt% polyacrylonitrile (PAN) solution ($MW = 150{,}000$ g/mol, $\eta = 800$ cP, $\sigma = 108$ S/cm) in dimethylformamide (DMF) using a multijet spinneret with three needles and the following conditions: voltage 29–30 kV, distance 15–19 cm, flow rate 2–3 ml/h, rotating disk at 300 rpm as collector, temperature 25°C, humidity 50–60%.

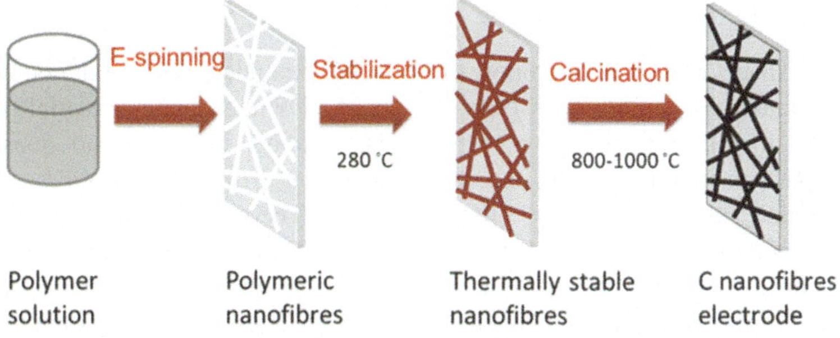

Figure 3.4 Schematic representation of the fabrication process of the electrospun CNF. souRce: Leitat.

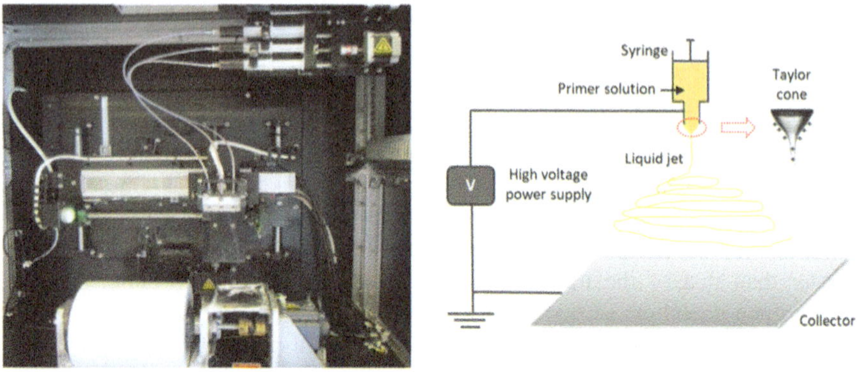

Figure 3.5 Photo of the device (left) and schema (right) of the electrospinning equipment. Source: Leitat.

The second step is the carbonization of the polymeric nanofibres to produce CNFs. In this process (Figure 3.6) the PAN nanofibres are first subjected to a thermal treatment above 200°C under O_2, in order to form a dense ladder-polymer that will prevent melting during subsequent carbonization. The carbonization takes place at temperatures between 600–1000°C under Ar in order to remove non-carbon elements in the form of different gases. During this process, fibres shrink in diameter losing approximately 60–85 wt% of their weight.

3.1.2.3 Collector materials

Within MIDES, two collector types based on carbonous materials were investigated for MDC applications: expanded graphite (bonded with fluoroplastic) and isostatic graphite (bonded with thermoset resin). Both concepts allow for the use of a high graphite content which is essential to achieve a higher electronic conductivity.

3.1.2.3.1 Flexible graphite sheets

Thin, flat sheets based on fluoropolymer-bonded expanded graphite are common collector materials in redox flow cells. The main advantage of this material is that it can be manufactured at low cost (semi-continuous process) and does not require post-machining. Sheets can easily be die-cut to the desired sizes and to accommodate the ports for flow cells (Figure 3.7). The use of a small amounts of thermoplastic binder based on fluoropolymers (ETFE, PVDF) allows for the manufacture of gas and liquid tight, thin sheets (0.6 mm) which generate a low weight footprint and can be conveniently die-cut. Sheets are very smooth and hence show a low contact resistance to the electrodes. The area-specific resistance (ASR) of these collectors is around 0.04 Ω cm^2.

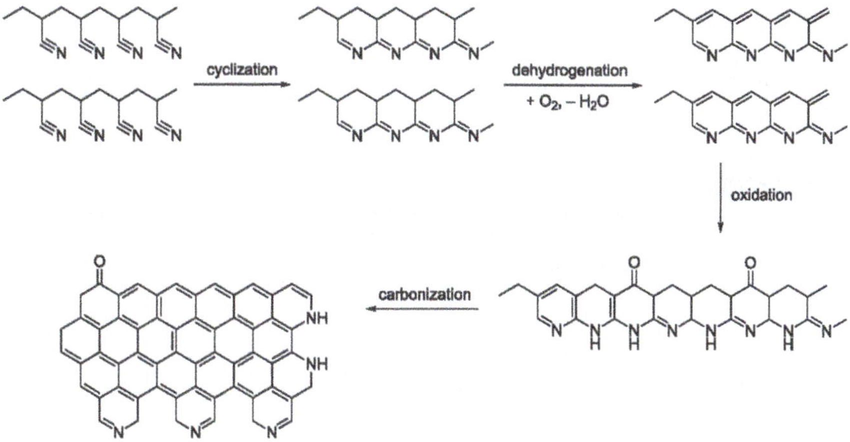

Figure 3.6 Scheme of CNF production from PAN nanofibers. Source: Gergin *et al.*, 2017.

3.1.2.3.2 Resin-impregnated isostatic graphite plates

Resin-impregnated, isostatically-pressed graphite is another route towards large-area collector sheets. A graphite block is formed by isostatic pressing of the carbon precursor and subsequent graphitization treatment. This block has a residual porosity and is therefore cut into slabs (dimensions close to the desired final size) and impregnated with thermoset resin. After curing, the material is machined to the final shape. The specific resistance of the collector (through-plane) was determined to be 265 mΩ cm, which is equivalent to an area-specific resistance (ASR) of 0.11 Ω cm^2 for a 4 mm thick plate (Figure 3.8).

3.1.3 Materials for electrode application in MDCs

3.1.3.1 Anode

A total of 25 carbon-based materials were tested regarding their capacity to allow electroactive biofilm growth, including unidirectional carbon fibres, woven and

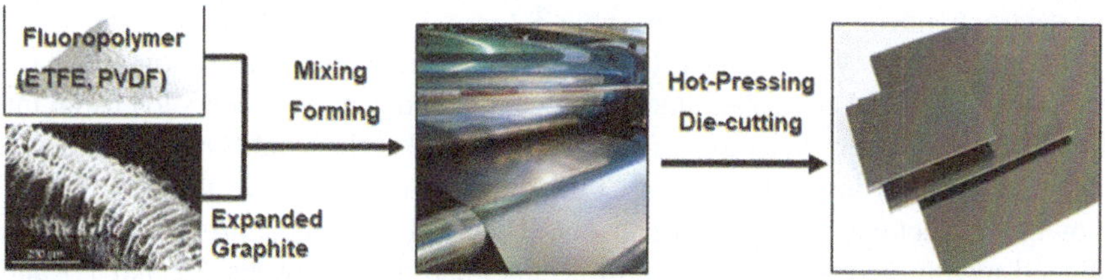

Figure 3.7 Manufacturing of collectors based on expanded graphite. Source: SGL.

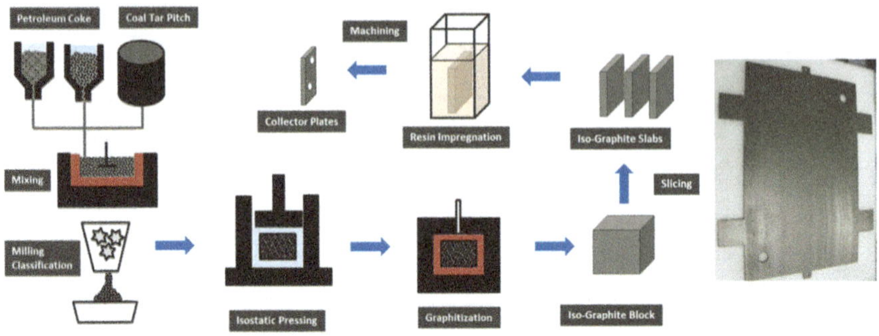

Figure 3.8 Manufacturing scheme of collectors based on resin-impregnated isostatic graphite (left) and collector prototype (right). Source: SGL.

non-woven fibres, and different types of carbon felts. Some of the materials were chemically and/or physically activated (see Table 3.2).

Surface activation is a common procedure for carbon electrodes which are intended for operation in aqueous environments. SGL's focus, as material provider, is on thermal surface oxidation by heating in air using reel-to-reel (R2R) processes. Thermal oxidation generates surface functional groups (e.g., –OH, –COOH, –C=O) which ensure wettability. In addition, the treatment increases the surface area (BET). Depending on the type of carbon fibre (only carbonized or after additional graphitization) different process parameters must be chosen.

Table 3.2 presents a selection of tested samples within MIDES to show the impact of the different post-treatments on the properties of the materials (BET surface, conductivity, hydrophobicity, and atomic composition). Also, the performance of the materials as anode electrodes (bioelectrochemical response), has been listed in Table 3.2, as very high/high/medium/low based on the biofilm growth chronoamperometry. Only the highest performing materials are shown in Figure 3.9. The best material tested so far seems to be GFD4.6-T700, since it exhibits the fastest colonization as well as a slightly higher current at the end with respect to the other materials. The lowest performing material as anode electrode in Figure 3.9 is KFD4.6EA which has been classified as medium bioelectrochemical performance in Table 3.2.

The chemical oxygen demand (COD), pH, conductivity and inlet flow rate have been monitored and controlled during the whole colonization process to ensure the reliability of different batches of tests carried out. The flow rates ranged between 100 and 200 ml/day, allowing a COD superior to 200 ppm, allowing proper growth of electroactive microorganisms. The feedstock medium consisted of a mineral medium containing 2.5 g/l of acetate, had a pH between 7.2 and 7.7 and conductivity between 9.5 and 11 mS/cm.

Table 3.2 Materials tested as anode materials, activation protocols, characteristics of the materials and qualitative bioelectrochemical performance.

Sample	Activation Method	Bio-Electrochemical Performance	Surface Area (m^2/g)	In Plane Resistance (Ohms/cm)	Transversal Resistance (Ohms)	Contact Angle (°)		Atomic Composition by XPS		
						Average (°)	Standard Deviation	%O At.	%N At.	%C At.
UD150	–	High	0.3	4.3	<0.1	high variation	*	**20.7**	**1.9**	**77.3**
UD150-HNO_3	Chemical (by nitric acid)	Medium	0.3	1.4	0.3	65.1	13.2	15.9	2.4	81.7
WF	–	Medium	0.3	4.0	1.8	0	0	16.7	3.9	79.4
WF-HNO_3	Chemical (by nitric acid)	Low	–	4.0	2.2	81.9	2.7	14.2	1.6	84.2
WF-T450	Thermal oxidation, 450°C 10 min	High	1.7	4.3	1.3	0	0	**11.5**	**3.8**	**84.7**
KFD4.6	–	Medium	0.4	2.2	2.7	0	0	8.6	4.1	87.3
KFD4.6-T300	Thermal oxidation, 300°C 10 min	High	29	4.1	3.9	0	0	**8.5**	**4.7**	**86.8**
GFD4.6	–	Medium	0.4	0.7	0.5	133.2	2.7	5	0.6	94.4
GFD4.6-HNO_3	Chemical (by nitric acid)	Low	–	–	–	–	–	–	–	–
GFD4.6-T700	Thermal oxidation, 700°C 30 min	High	3.9	0.5	0.4	126.0	3.3	9.4	0.9	89.7
GFD4.6-CO_2-T950	CO_2-thermal	Low	0.5	0.4	0.3	115.9	9.5	**3.9**	0.7	95.4

UD: unidirectional fibres; WF: Woven fabrics; KFD: carbonized carbon felt; GFD: graphitized carbon felt; XPS: X-ray photoelectron spectrometry.

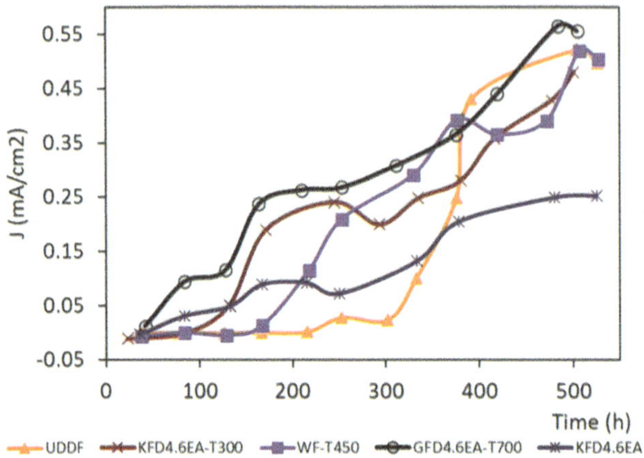

Figure 3.9 Biofilm growth chronoamperometry of the best performing materials. Source: Leitat.

Table 3.2 shows the specific surface area measured with BET. The results show that all materials have high and sufficient surface area for MDC applications. The highest increase in specific surface area after different treatments is obtained when thermal activation is used. Among the materials studied, KFD4.6 showed the highest specific surface area when treated at 300°C in air for 10 min. Although all the materials have sufficiently large surface area to be used as MDC biotic anodes, a relationship can be observed between the increase of surface area and the performance of the material as an anode.

The conductivities of the materials were checked to ensure that they can be used in the MDC application. Thus, in terms of conductivity all the tested materials seem to be appropriate. Contact angle measurements were taken to study the wettability of the carbon electrodes, and also for the activated/treated materials. The hydrophilicity properties shed light on the adhesion of the microorganisms to the electrode. As the bacteria are present in the solution, a better electrode-liquid interaction implies an improvement in the bacterial adhesion. In general terms, carbon felt presents lower resistance values than other carbon fibres (UD and WF). Regarding carbon UD fibres, the most significant result is the difference in the resistance when the current flows along the axial direction, with respect to the conductivity in the perpendicular direction, which decreased to values of hundreds of mΩ. Treating the carbon fibres with the nitric acid attack increased the conductivity to values close to 1 Ω.

For carbon felt samples, the contact angle (initially 145°) decreases after both heat (126°) and nitric acid (125°) activation, which implies an increase of the hydrophilicity of the electrode. In the case of the carbon fibres, it was found that the drop wetted the sample completely, probably due to the distance between the fibres and the capillary forces that may be caused by these and the substrate. For the treated fibres, a similar behaviour was observed: the drop either wetted completely or remained initially with a contact angle of 65°. Contact angle results for the carbon fibres were not reproducible enough to be considered. The contact angle of carbon felt KFA6EA, KFD4.6EA, KFD2.5 materials was 0 due the water drop being absorbed in the material indicating a good wettability (hydrophilic materials).

The surface chemistry of the anode materials is one important factor for the formation of a biofilm. Therefore, XPS was used to check the chemical composition of the anode materials. This characterization was performed for the base materials, as well as for materials after activation treatment. As mentioned previously, thermal or acid oxidation generates surface functional groups (–OH, –COOH, –C=O), which improves the wettability of the material. The overview spectrums from all the samples reveal the presence of C, O and N. No other elements were detected. This is because all investigated materials are based on a polyacrylonitrile precursor. By the carbonization treatment this precursor is transformed into carbon with increasing carbon content with increasing carbonization temperature.

The untreated base materials show a carbon content between 77.3 and 94.4% and oxygen content between 5.0 and 20.7%. The highest carbon and lowest oxygen contents were found for the GFD, which is the material which was treated at the highest graphitization temperature. The UD material exhibits the highest oxygen content. This is not only based on the carbon fibres, but rather on the organic sizing which coats the fibres. Such a sizing is also present for the WF, which also exhibits a high oxygen content.

By applying different activation treatments, the surface composition was altered, but a significant increase in oxygen content was only observed for the GFD4.6-700 (from 5% to 9.4%). This surface modification resulted in a strong improvement in cell performance. For KFD4.6, just a slightly increase in nitrogen content occurred, but nevertheless the activated material exhibited a better performance in MFC application.

For UD and WF, the two materials consisting of carbon fibres with organic sizing, conversely a decrease in oxygen occurred by applying the activation treatment. This can be traced back to the fact that the activation treatment eliminated the oxygen-rich surface coating and exposed more of the carbon fibre surface. However, the sizing on the carbon fibres of the UD seems to be advantageous for the colonization process, since the untreated UD showed a very good cell performance. For the WF, however, the thermal treatment improved the cell performance even though the oxygen content was lowered. Most likely, the sizing on this woven fabric is not as suitable for microbial colonization.

Unfortunately, based on this XPS study, no clear relationship between surface chemistry and inoculation performance could be identified. Additionally, of course, the surface area and surface structure of the fibres are also important for the formation of the biofilm. The surface topography might even play the key role, if the surface chemistry basically fits, as it is provided by carbon materials with a certain amount of oxygen.

Scanning electron microscope (SEM) imaging was performed for the different carbon-based materials that were tested as anodes. The aim was to observe if the materials suffered any topographical modification as a consequence of the treatments carried out. Though in general terms there were no significant changes between the different treated samples with respect to the untreated ones, there are two cases that are worth mentioning. On the one hand, the thermally treated WFCF material presents a smooth change in the roughness with respect to the untreated sample, and on the other hand, the GFA6EA material treated with nitric acid also presents a change in the surface roughness. The diameter of the microfibres observed was in the range of 7–10 μm for all cases. The SEM images of samples obtained before the biofilm growth test, show the microorganisms adhered on the carbon fibres. In these images the quantity of biofilm can be qualitatively determined, and this is in concordance with electrochemical characterization of inoculated anodes. The best materials detected in the chronoamperometry characterization (Figure 3.9) show a higher biofilm colonization of the surface in the SEM images (Figure 3.10). This is the case for GFD4.6EA-T700, UDCF150 and KFD4.6EA-T300. On the contrary, the

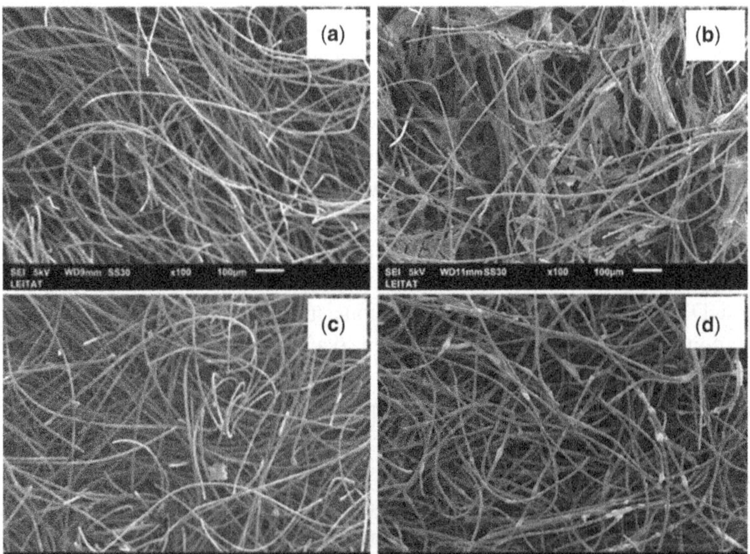

Figure 3.10 SEM images of inoculated carbon material (GFA4.6EA-T700) and material with low biofilm growth (GFD4.6EA-T950-CO_2). Left image (a and c) belongs to raw materials and right image (b and d) belongs to materials after biofilm growing experiments. Source: Leitat.

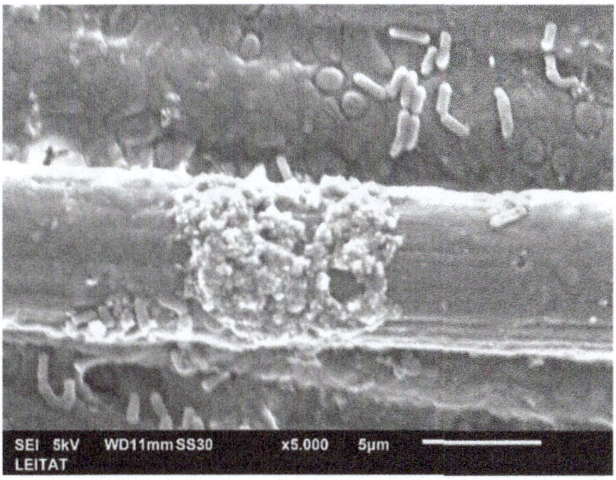

Figure 3.11 SEM image of grown electroactive biofilm onto unidirectional fibres UD150. Source: Leitat.

materials that generate a low current in chronoamperometry, indicating a low rate of biofilm growth, show a low amount of biofilm in the SEM images (UDCF150-HNO_3, GFD4.6EA-T950-CO_2, and GFD4.6EA-HNO_3).

Several samples for SEM were taken from all carbon materials tested as anodes to assess biofilm formation on the surface of the electrode. In all of the samples with electroactivity, biofilm was observed (data not shown). Figure 3.11 shows a biofilm growing on unidirectional fibres. A homogeneous layer covering the fibres can be observed. This layer corresponds to the typical conglomeration of polymers, proteins and polysaccharides that constitute a biofilm. Also, in the image, several microorganisms can be identified, some of them within the biofilm matrix and some others on the surface of it.

DNA was extracted from different anode samples after inoculation with electroactive biofilm. It was observed that the material influenced the final anodic biofilm composition (Figure 3.12). However, no clear relationship from physico-chemical parameters and trends in microbial population could be established.

3.1.3.2 Liquid cathode

Liquid cathodes based on carbon felts were investigated in flow-battery-type cells separated by a Nafion® N-115 membrane (125 μm). Both half cells were fed with identical electrolyte (25 mM ferricyanide/25 mM ferrocyanide as a redox couple dissolved in water with different salinity) from a single container in batch recirculation mode at very high flow rates. This straightforward method (Schweiss *et al.*, 2020) allows for a dissection of different resistances and thus allows calculation of the expected cathode losses in an MDC.

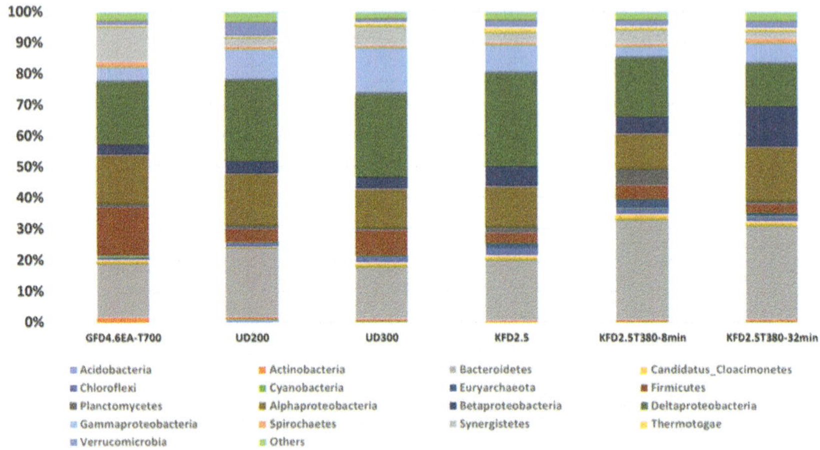

Figure 3.12 Microbial population grown on different anode materials, characterized by pyrosequencing. Source: Leitat.

It is verified that carbon felt cathodes are highly effective and for the current densities of MDCs the electrode-related losses amount to only a few millivolts (Table 3.3). Hence, the potential of the cathodes during MDC operation will only depend on the degree of conversion of oxidant species (e.g., Fe^{3+}) according to the Nernst equation (equation 3.1)

$$E_{\text{cath}}(V) = 0.77 + 0.026 \cdot \ln \frac{c_{\text{Fe}^{3+}}}{c_{\text{Fe}^{2+}}} \tag{3.1}$$

Comparing different fibre treatments, is has been verified that graphitization of the fibre does not provide a substantial benefit for a carbonized fabric (e.g., felts of KFD type).

Table 3.3 Effective resistance of carbon felt (KFD type) electrodes using electrolytes containing 50 mM redox species and different salinity levels.

Electrolyte Salinity (g/l)	Electrode Resistance (Ω cm^2)		
	Ohmic	Kinetic	Total
5	0.85	1.75	2.60
15	0.77	1.50	2.27
35	0.63	0.99	1.62

Resistances were calculated from polarization resistance and impedance of double-half cells (Schweiss, 2020).

Screening tests of different carbon felt (identical to those used as anodes) with Fe^{3+} indicated that the kinetics of iron or ferrocyanide species is fast and virtually independent of the surface chemistry of the felt. The only difference is given by the electronic conductivity which determines the portion of electronic resistance of the porous electrode. Table 3.4 shows the measured cell resistance for different cells (50 mM ferrocyanide in phosphate-buffered saline medium (PBS), total salinity 35 g/l) and the cell impedance (which only has contributions from the two electrodes, the electrolyte and the Nafion 115 membrane) as well as the apparent rate constant for charge transfer (k0). Surface treatment of the graphitic felts results in a lower cell resistance due to a higher wettability/utilization of the surface area of the felt.

All the felts studied should therefore be suitable for a MDC liquid cathode since the ohmic (IR) losses of the electrode could be tolerated. Additionally, the IR losses could be mitigated by making the felt materials thinner. Also, trials towards a GFD 2 and 2.5 mm were conducted successfully.

The electrochemical behaviour of SGL materials as cathodes was evaluated in an electrochemical device using $K_3[Fe(CN_6)]$ 0.06 M as the catholyte solution. The reduction reaction at the cathode compartment is as follows:

$$[Fe(CN_6)]^{3-}(ac) + e^- \rightarrow [Fe(CN_6)]^{4-}(ac) \rightarrow E^0 = +0.36\,V$$

The rationale of the use of this iron salt as a catholyte relies on its fast kinetics, low cost compared to other metal complexes and the possibility for it to be regenerated biologically. On the other hand, it may cause membrane damage in long-term experiments due to slightly corrosive media, precipitates on the cathode electrode surface may occur, and in the case of leakage cross contamination of the saline and anolyte stream may take place. Table 3.5 summarizes all SGL materials tested.

The electrochemical behaviour of the materials provided by SGL was studied by means of polarization curves (Figure 3.13). These curves represent the operative behaviour of the cathode in the microbial electrochemical device.

The best materials to be used as electrodes for liquid cathodes are PV15-GFA 4.6 EA (700C/30 min) and PV15-KFD 4.6 EA. MDCs employing liquid

Table 3.4 Effective cell resistance and impedance of different cathode materials.

	KFA6	KFD4.6	GFA6	GFA6 (450°C/1 h)	GFD4.6
Cell resistance (Ω cm^2)	6.7	5.1	6.6	5.4	4.9
Cell impedance (Ω cm^2)	4.6	3.1	4.6	3.1	2.6
k0 in ferrocyanide (10^{-2}cmgeos^{-1})	0.83	0.82	1.04	1.05	0.91

Table 3.5 SGL materials tested as cathode using liquid iron salt catholyte.

Cathode Collector	Cathode Carbon Felt
PV15	GFA 6 EA (450°C/1 h)
	GFD 4.6
	GFA 6 EA
	GFA 4.6 EA (700°C/30 min)
	KFD 4.6 EA
	KFA 6 EA

cathodes (e.g., Fe mediators) could be constructed based on symmetric carbon felt electrodes. The reduction of thickness to 2 or 2.5 mm is favourable due to lower net resistance and cost.

3.1.3.3 Air-cathode

The oxygen reduction reaction (ORR) for different air-cathode materials was investigated by cyclic voltammetry (CV) and chronopotentiometry in PBS buffer. The CV shows the ORR behaviour of the air-cathodes and the potentials where the reaction is produced.

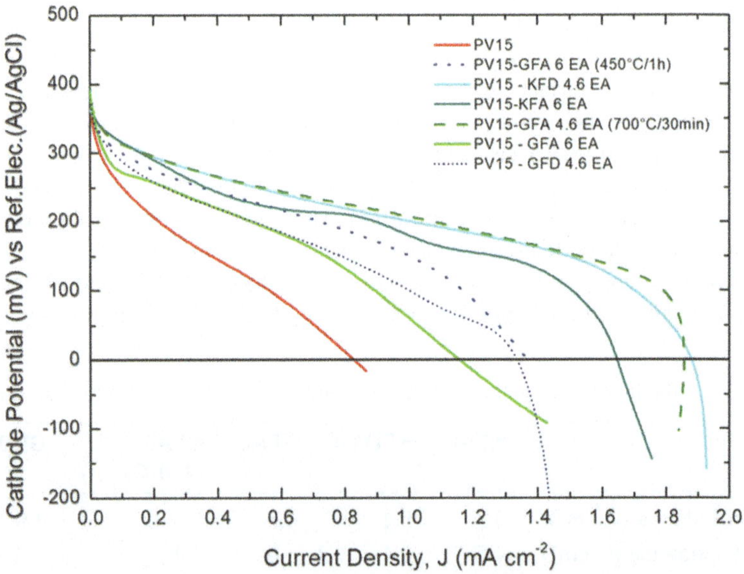

Figure 3.13 Behaviour of SGL materials in cathodic liquid conditions. Source: IMDEA.

Even at the very low current densities of around 1 mA/cm^2 which are typical in MDCs, the potential drops by several hundred mV for all types due to the high overpotentials of the ORR (which is exacerbated by the low proton concentration). For example, at 1 mA/cm^2, the high surface area manganese dioxide based air-cathode showed a higher potential than the reference platinum based air-cathode (MnO_2 + 120 mV, C/Pt – 40 mV, both vs. Ag/AgCl). This confirms the suitability of manganese dioxide as an efficient non-PGM ORR electrocatalyst at neutral pH (Roche & Scott, 2009), but also iron subgroup metals represent a good alternative to platinum (Bosch-Jimenez *et al.*, 2017).

3.1.3.3.1 Development and results of free-standing Fe/Co doped carbon nanofibres

Pristine carbon nanofibres can offer good ORR activity due to their high surface area. In order to improve the ORR activity, Co and Fe nanoparticles were incorporated in the carbon nanofibre matrix by means of a precursor route. For this purpose, different quantities of the metal acetate $Co(COOCH_3)_2$, Fe $(COOCH_3)_2$) were mixed with the polymer/s solution and electrospun afterwards. After the thermal treatment, metal nanoparticles (Co or Fe) embedded in the matrix of carbon nanofibres were obtained. It is supposed that the presence of the acetate together with the high carbon content are the responsible factors for the reduction of the metal salt to the metal (0).

The most important tool to screen the materials prepared was the ORR evaluation using abiotic electrochemical techniques. ORR performance was the selection criteria to optimize the catalyst content (range: 1 to 6 mmol), polymer/s concentration in the dope solution, composition (PAN or PAN/ polyvinylpyrrolidone-PVP mixture), etc. In addition, different thermal treatments with a final temperature from 800 to 1200°C, were carefully studied. A total of 40 materials were prepared during MIDES but only those complying with the mechanical properties for MDC usage were evaluated.

In order to simplify the discussion about the more relevant results obtained in the preparation and performance of materials, a selection of materials based on Fe catalyst and Co catalyst is presented.

The composition and preparation details for the Fe doped materials mentioned above are included in Table 3.6. Fe-doped CNF had to be optimized in terms of polymer precursor composition, since very poor mechanical properties were observed and limited charge of catalyst per weight was achieved when PAN was used as the precursor. PVP, a well-known material used in several applications as a dispersant, was employed for improving the mechanical properties and performance of ORR. Note that lower pyrolysis temperatures than in Co-doped CNF were needed because PVP is mostly pyrolysed at that temperature creating porosity.

Table 3.6 Fe doped CNF composition.

Entry	Sample Labelling	Catalyst Precursor (mmol)	Concentration Pre-Polymer (% wt)	Thermal Treatment Tf (°C)*	ORR Performance
1	Fe3–900	3	12 (+PVP)	900	Good
2	Fe2–800	2	12 (+PVP)	800	Good
3	Fe3–1000	3	12 (+PVP)	1000	Bad
4	Fe3–800	3	12 (+PVP)	800	Bad

*Tf: final temperature.

Figure 3.14 shows the ORR test of the 2 best and 2 of the worst iron doped carbon nanofibres of the 10 materials developed and tested. The best candidate as air-cathode considering the iron based carbon nanofibres was Fe3-900 as it exhibited higher reductive current at more positive voltages (0 to 0.3 V), as detailed in the reduction scan of cyclic voltammetry (CV) graph. The high open circuit potentials (OCPs) of different iron-based materials indicate that all the carbon nanofibre materials have the presence of iron as a catalyst. Also, the OCP tendency is aligned with the ORR performance shown in the CV graph. But, to evaluate the best performance materials in terms of ORR, the CV graph is more relevant than the OPC graph.

In the following paragraphs, details about Co-doped CNFs are presented. The composition and details of preparation are included in Table 3.7. In this table the composition of the precursor in the solution to obtain the CNFs is shown. It can be observed that PVP can be used but it is not completely necessary to obtain

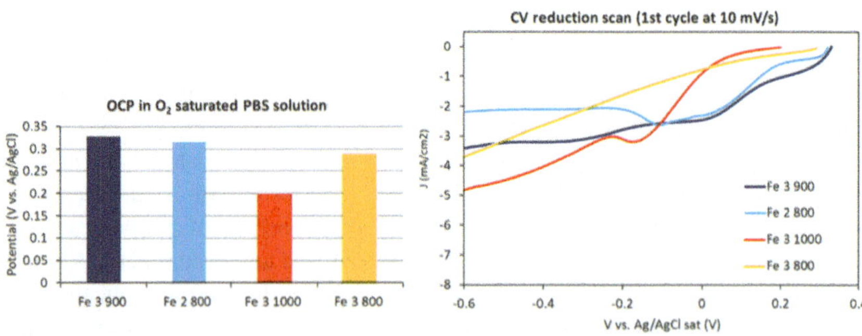

Figure 3.14 Open circuit potential (OCP) (left) and reduction scan of cyclic voltammetry (right) of two best and two worst iron doped CNFs. The tests have been carried out in O_2 saturated PBS. Source: Leitat.

Table 3.7 Co doped CNF composition.

Entry	Sample Labelling	Catalyst Precursor (mmol)	Concentration Pre-Polymer (% wt)	Thermal Treatment Tf (°C)	ORR Performance
1	Co5–1000-PVP	5	12 (+PVP)	1000	Good
2	Co5–1000-P10	5	10	1000	Good
3	Co6–1000-P10	6	10	1000	Bad
4	Co5–1000-P12	5	12	1000	Bad

CNFs with good performance ORR (entry 1 and 2). Co-doped CNFs admit higher temperatures (1000°C) than Fe-doped CNFs and higher content (up to 5 mmol) than Fe-doped ones (3 mmol). Higher content of Co acetate has a negative effect on ORR.

Figure 3.15 shows the ORR test of the two best and two worst cobalt doped CNFs of the nine materials developed and tested. The best candidate as an air-cathode considering the cobalt based CNFs, is Co5-1000* as it exhibits a higher reductive current at more positive voltages (0 to 0.25 V) in the reduction scan of cyclic voltammetry (CV) graph. The OCPs of cobalt material are lower than the iron ones but these are also high. The high values indicate, as in the case of iron-based materials, that the CNFs contain the metal catalyst. The OCP difference between the best and worst cobalt doped material is approximately 100 mV, and the OCP tendency is aligned with the ORR performance shown in CV as well.

Figure 3.16 shows the reduction scan of CV and chronopotentiometry of the best material of iron doped CNFs and best material of cobalt doped CNFs. The iron

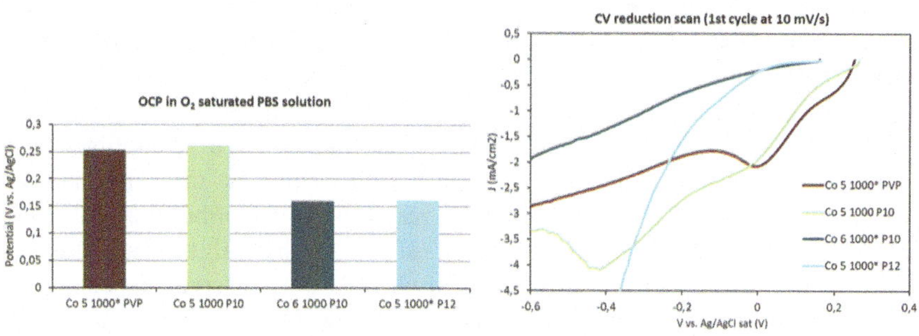

Figure 3.15 Open circuit potential (OCP) (left) and reduction scan of cyclic voltammetry (right) of two best and two worst iron doped CNFs. The tests have been carried out in O_2 saturated PBS. Source: Leitat.

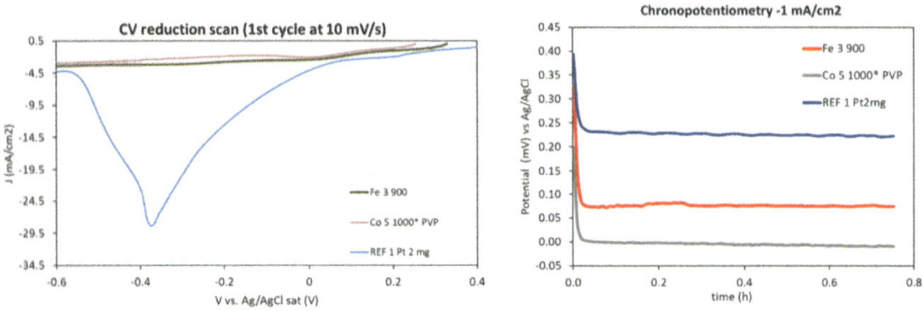

Figure 3.16 Reduction scan of cyclic voltammetry (left) and chronopotentiometry (right) of the best cobalt and iron doped CNFs materials with platinum coated carbon cloth as reference material. Source: Leitat.

doped CNFs have higher ORR performance than cobalt doped CNFs, as the reduction current in the CV graph is higher, and in the chronopotentiometry tests, applying a reductive current of 1 mA/cm^2, the material potential is maintained at a more positive potential (at 80 mV for Fe doped CNFs and 0 mV vs. Ag/AgCl for Co doped CNFs). Thus, in terms of ORR performance the Fe3-900 is the best material.

A comparison of the materials developed in MIDES with respect to platinum coated carbon cloth materials (as state-of-the-art reference) was made. In the CV graph, the reductive current of the reference materials was higher indicating a higher performance, however at working potentials of the air-cathode (0.1 to 0.25 V, working the cell close to maximum power point), the current exhibited by the reference material was still higher than Co and Fe doped CNFs but was not very different. The chronopotentiometry graph, also indicated that the reference material had higher ORR performance with respect to Co and Fe doped CNFs, as the potential was maintained at a more positive potential (0.23 V vs. Ag/AgCl). The Fe3-900 had lower performance than the platinum-based material, but the results and the comparison with platinum material, demonstrate that the performance of Fe3-900 is sufficient to use this material as an air-cathode in an MDC reactor.

Microscope images of the different doped CNFs confirmed the presence of the nanoparticles embedded into the carbon nanofibre surface. Nanofibres and nanoparticle diameters of Fe- and Co-doped CNFs could be measured and are included in Tables 3.8 and 3.9. Usually, a reduction in nanofibre diameter from 400–700 nm to 200–400 nm is observed after the thermal treatment of the electrospun fibres. A variable weight loss of 60–85% was observed depending on Tf, catalyst used, and concentration. Standard deviation for the samples was included to indicate how homogeneous the tested CNFs were in terms of nanoparticle (NP) distribution, CNFs range of diameters, etc.

A few examples of the best materials can be found in Figure 3.17.

Table 3.8 Fe doped CNF composition, performance and CNF and NP sizes in nm (standard deviation).

Sample Labelling	Catalyst Precursor (mmol)	Concentration Pre-Polymer (% wt)	Thermal Treatment Tf (°C)	Yield (%)	ORR Performance	(a) CNF Diameter (+SD) (b) NP Diameter (+SD) (nm)
Fe3–900	3	12 (+PVP)	900	17	Good	(a) 422 (56) (b) 14.3 (5)
Fe2–800	2	12 (+PVP)	800	28	Good	(a) 267 (37) (b) 13.5 (3.5)
Fe3–1000	3	12 (+PVP)	1000	21	Bad	(a) 437 (48) (b) 20.9 (14.2)
Fe3–800	3	12 (+PVP)	800	32	Bad	(a) 544 (51) (b) 15.0 (5.1)

Table 3.9 Co doped CNF composition, performance and CNF and NP sizes in nm (standard deviation).

Sample Labelling	Catalyst Precursor (mmol)	Concentration Pre-Polymer (% wt)	Thermal Treatment Tf (°C)	Yield (%)	ORR Performance	(a) CNF Piameter (+SD) (b) NP Piameter (+SD) (nm)
Co5–1000-PVP	5	12 (3%PVP)	1000	21	Good	(a) 260 (88) (b) 35.8 (8.1)
Co5–1000-P10	5	10	1000	35	Good	(a) 318 (34) (b) 13.6 (3.9)
Co6–1000-P10	6	10	1000	9	Bad	(a) 374 (52) (b) 22.6 (7.4)
Co5–1000-P12	5	12	1000	34	Bad	(a) 547 (66) (b) 57.7 (31.5)

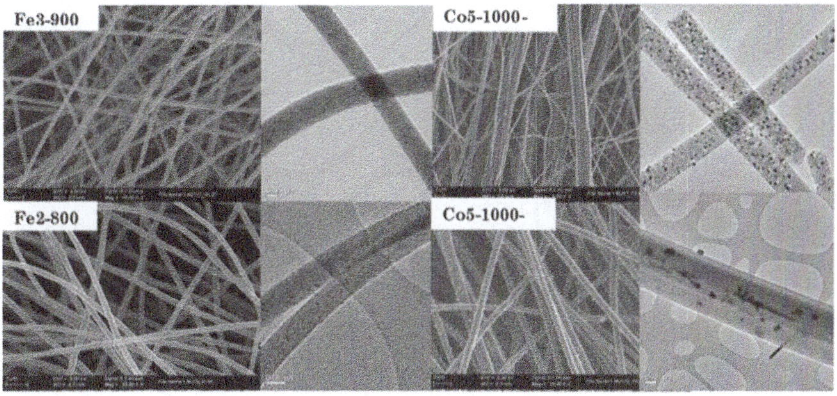

Figure 3.17 High-resolution scanning (HR-SEM) and transmission (HR-TEM) electron microscopy and of Fe-doped CNFs (left) and Co-doped CNFs (right). Source: Leitat.

As mentioned before, the thermal treatment final temperature has a significant influence in the CNF morphology and nanoparticle size and distribution along the NF, as can be seen for the case of Fe-doped CNFs in Figure 3.18.

3.1.3.3.2 Gas diffusion electrodes (GDEs) based on MnO_2

For this air-cathode material ORR tests were performed in two ways using PBS with a salinity of 35 g/l, as follows:

(a) A single three-electrode cell with a cathode immersed into *oxygen-saturated solution* with oxygen bubbling maintained during the measurements. Ag/AgCl and Pt were reference and counter electrodes, respectively. This was mainly used for catalyst screening purposes.
(b) An air breathing cell (Figure 3.19): cylindrical cells containing, with a breathing cathode, an Ag/AgCl reference electrode with a Luggin capillary and a large area Pt-coated titanium sheet counter electrode were

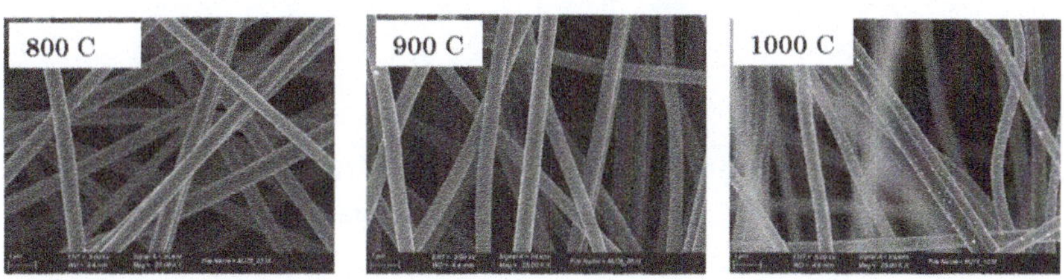

Figure 3.18 HR-SEM of the Fe-doped (3 mmol) CNFs treated at different temperatures. Source: Leitat.

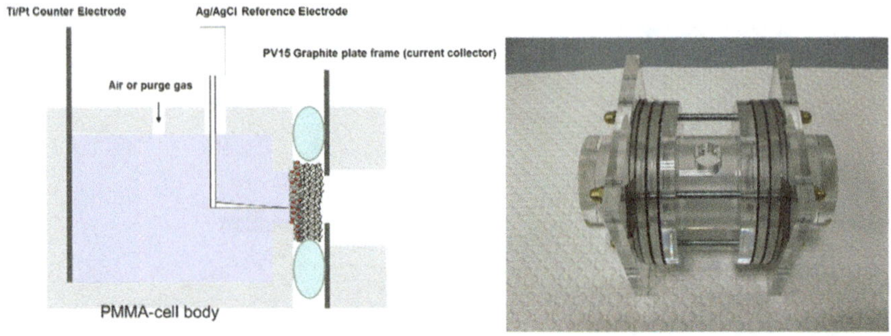

Figure 3.19 Air-breathing cell constructed in polymethyl methacrylate (PMMA). Source: SGL.

used. This method was used for characterization of pre-selected candidates under conditions which are nearer to the final application. The cathode was in contact with PV15-type graphite current collector frames.

Figure 3.20 shows a comparison of different air-cathodes in oxygen-saturated solutions and in cells with a breathing cathode.

For the breathing cathode, the GDE based on MnO_2 catalyst shows ORR activity, although it is lower than that of the Pt-based air-cathode.

3.1.3.4 Electrochemical characterization of best air-cathode materials at MDC operational conditions

Both air-cathode strategies were further tested for their application in MDCs. In a first approach, an electrochemical characterization was undertaken, which was followed by a desalination trial.

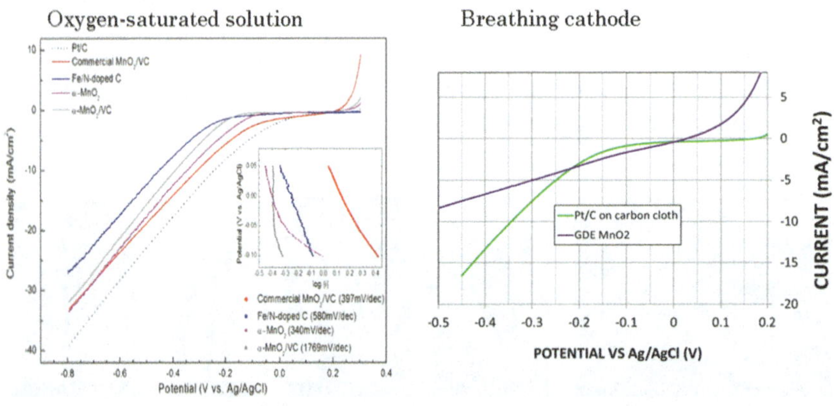

Figure 3.20 Results for gas diffusion electrode (GDE) performance based on MnO_2. Source: SGL.

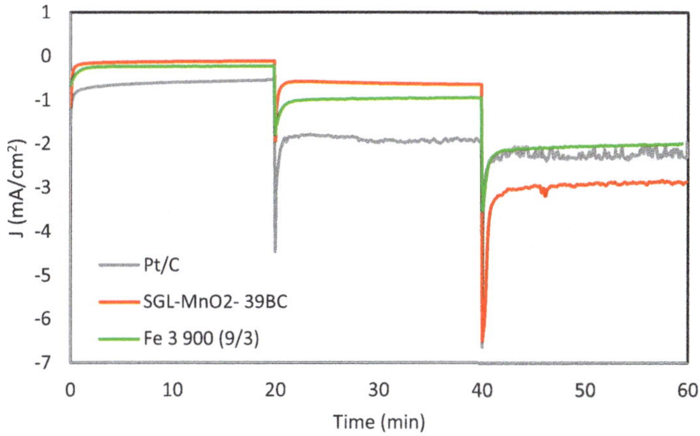

Figure 3.21 Constant voltage technique, bubbling O_2 and stirring at 0.1, −0.1 and −0.3 V vs. Ag/agCl for 20 minutes each. Working electrode surface of 0.28 cm^2; PBS 100 mM; pH = 7; Medium conductivity (k) of 13 mS/cm. Source: Borràs *et al.*, 2021.

Figure 3.21 shows the constant voltage experiment, where three different voltages (+0.1, −0.1 and −0.3 V vs. Ag/AgCl) were applied. Higher reduction current density for Fe3-900 (9/3) at 0.1 and −0.1 V vs. Ag/AgCl was obtained compared to SGL-MnO_2-39BC. However, at −0.3 V, SGL-MnO_2-39BC produced a higher current than Fe3-900 (9/3), which indicates that when a sufficiently negative potential is applied, SGL-MnO_2-39BC is more effective with regard to ORR. Hence when sufficient overpotential is applied, SGL-MnO_2-39BC will be a better cathode for MDCs. Similar behaviour has been reported by Gyenge and Drillet (2011). At low overpotential, a platinum-based cathode material works better than the other two, however the difference is not very significant. But at the higher overpotential tested, the current of SGL-MnO_2-39BC was higher than the platinum material indicating higher ORR performance and the Fe3-900 (9/3) had similar current output indicating similar ORR at this potential.

Finally, air-cathode materials were tested in an operational air-cathode MDC performing desalination (Figure 3.22). Due to the MDC reactor being operated under very low external resistance (2.5 Ω), close to short-circuit conditions, the cathodic potentials are negative. Fe3-900 (9/3) performed with high stability as the electrochemical potential remained almost constant in comparison with the SGL-MnO_2-39BC cathode. As observed in abiotic ORR characterization, SGL-MnO_2-39BC has higher performance than Fe3-900 (9/3) at this working potential. This produces a higher current in the MDC reactor and also a higher desalination rate, which is emphasized during the last hours of the SGL-MnO_2-39BC air-cathode-MDC experiment, when the cathode performance

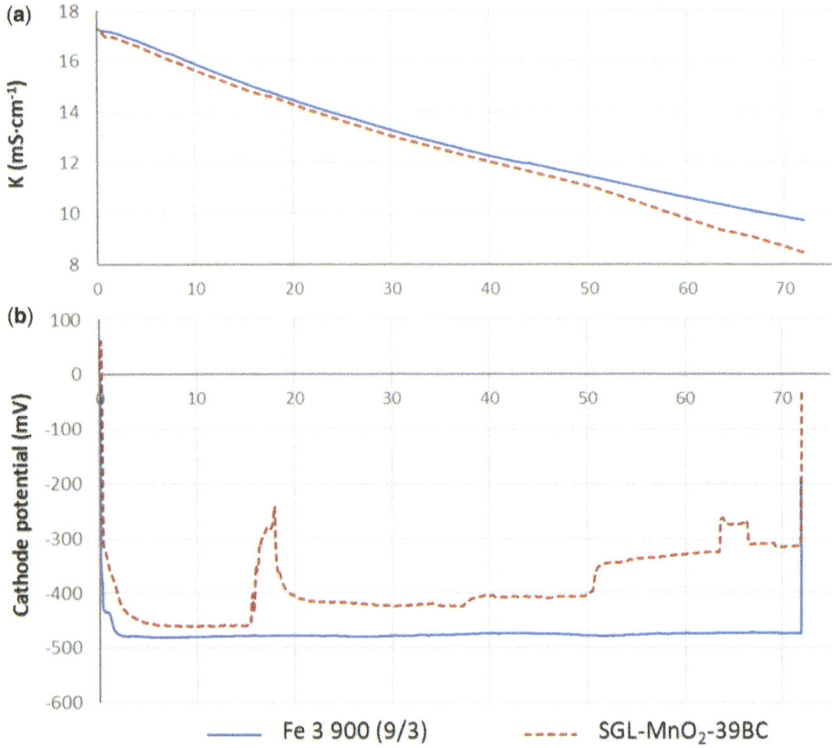

Figure 3.22 MDC operated at fixed external resistance of 2.5 Ω with Fe3-900 (9/3) and SGL-MnO_2-39BC based air-cathodes. (a) conductivity evolution of saline chamber, (b) cathode potential evolution. Source: Borràs *et al.*, 2021.

increases, increasing the current and the desalination rate, as well. The high overpotential that both cathode materials exhibited during the MDC experiment was due to the operating conditions being very close to short-circuit conditions.

3.1.3.5 Upscaling air-cathodes based on iron doped carbon nanofibres

An attempt to upscale the air-cathode MDC was made in the frame of the MIDES project. The upscaled cathode was constructed with CNFs doped with iron (Fe-3-900 material). To decrease the internal resistance of the cathode, unidirectional carbon fibres and expanded graphite (both provided by SGL) were added to the CNFs. Two types of TYVEK® textile were placed at each face of the cathode. On one side the TYVEK® acts as a dielectric separator, while allowing humidity pass and proton transport. On the other side the TYVEK® acts

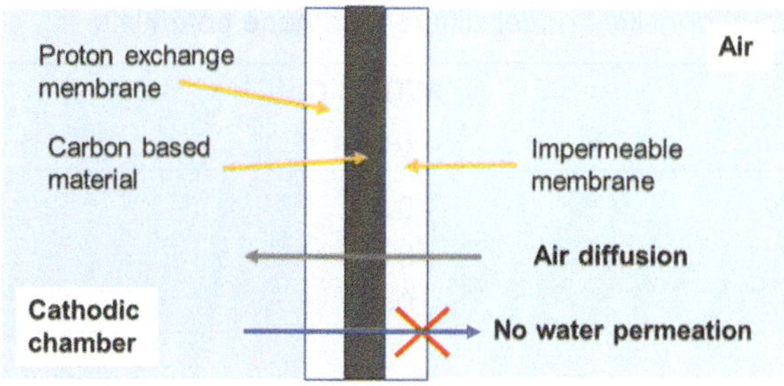

Figure 3.23 Air-cathode scheme in the air-cathode MDC. Source: Leitat.

as a water impermeable and air diffusion membrane (Figure 3.23). The final active area of the cathode was 580 cm^2.

The upscaled air-cathode was initially characterized using an air-cathode microbial fuel cell. The performance of this air-cathode was compared with an operating air-cathode fabricated using commercial catalyst (also platinum-free catalyst). To determine the ORR performance, the following characterizations were carried out:

- Open circuit potential of air-cathodes (OCP cat): measured with three-electrode configuration, using an Ag/AgCl reference electrode. The cell was in rest operation for 3 hours before measuring the OCP cat.
- Cathode oxygen reduction current under cathode polarization at different potentials (CstV): the currents at different CstV were measured with three-electrode configuration, using an Ag/AgCl reference electrode. The cathode was polarized at three potentials (100, 50 and 0 mV vs. Ag/AgCl) and the output intensities recorded.
- Cathode potential at different constant current discharge (CstC): measured with three-electrode configuration, using an Ag/AgCl reference electrode. Three different negative currents (reduction current) were applied and the resulting cathode potential recorded.
- Polarization curve (PC): performed by applying external resistance from 1000 to 5 Ω, with a stabilization time of 30 minutes at each resistance. The PC was performed with two-electrode configuration cells.

The OCP_{cat} was +200 mV vs. Ag/AgCl and the OCP_{cat} of the reference cathode was −50 mV vs. Ag/AgCl. The higher OCP of the cathode material developed in MIDES indicates lower ORR overpotential.

The characterization of reduction currents at three different potentials (+100, +50, and 0 mV vs. Ag/AgCl) indicates the kinetics of the cathode ORR. The more negative the current is, the higher the cathode ORR rate is, allowing a better performance of the whole MDC cell. As Table 3.10 indicates, the pre-pilot cathode allowed negative currents at all three applied potentials, indicating good kinetics for ORR.

Table 3.10 Current densities under different cathode potentials.

	MIDES Cat J (A/m^2)	Ref Cat J (A/m^2)
CstV +100 mV	−0.06	+0.04
CstV +50 mV	−0.15	+0.03
CstV −50 mV	−0.33	−0.01

In addition, the air-cathodes were tested applying negative current and measuring the steady state potential where the air-cathodes stabilize. Air-cathodes with proper ORR performance are able to accept high current densities with a small shift of cathode potential to negative redox potential. In contrast, air-cathodes with poor performance need a high shift of potential to negative redox potentials to carry out the ORR. Table 3.11 shows the steady state potentials of both cathodes, and it can be observed that the MIDES cathode are more positive than the reference cathode for both applied currents. These results also show that the oxygen reduction reaction performance of the MIDES cathode is better than that of the reference cathode.

The polarization curves of the air-cathode MFC (Figure 3.24), one constructed with the MIDES cathode and the other with the reference cathode, show that both cells have proper operation. The MFC containing the MIDES cathode enabled significantly higher performance. The anode performances were equivalent (data not shown here). Thus, the higher performance of the MFC is due to the higher performance of the MIDES air-cathode with respect to the reference cathode. This was in concordance with all previous electrochemical characterizations.

In conclusion, the scaling up of an air-cathode material, fabricated with iron doped CNFs, was successfully validated in an air-cathode MFC to determine the oxygen reduction reaction performance. Results were compared with a reference cathode fabricated with commercial materials. The cathode material developed in MIDES showed higher OCP_{cat} potential, indicating lower reaction overpotentials. Also, constant voltage, constant current and polarization curve characterization showed better performance of the MIDES cathode. Thus, the overall ORR performance of this air-cathode opens the door for future trials of air-cathode MDCs at pre-pilot scale.

Table 3.11 Cathode potential at different constant current values.

	MIDES Cat P Cat (mV)	Ref Cat P Cat (mV)
CstC −15 mA	+20	−140
CstC −30 mA	−150	−210

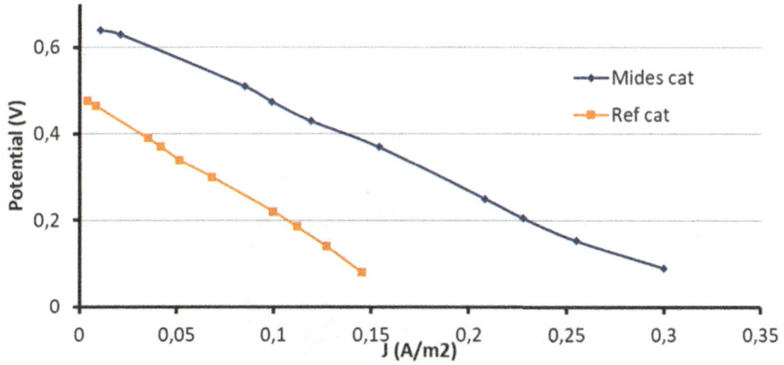

Figure 3.24 Polarization curves of air-cathode microbial fuel cell containing MIDES air-cathode (blue) and reference cathode (orange). Source: Leitat.

3.2 MEMBRANES

3.2.1 IEM general specifications

Ion-exchange membranes (IEMs) are used in a number of processes which are rather different in their basic concept, their practical application, and their technical relevance. However, in all these processes membranes carrying electrical charges are used to control the transport of ionic species and to separate them selectively from a mixture with neutral components (Strathmann, 2004). One of these technologies is the microbial desalination cell (MDC), see Figure 3.25. Here, a potential gradient is created across the anode and cathode chambers, and the anions (such as Cl^-, NO_2^-, HCO_3^- and SO_4^{2-}) migrate from the saline chamber across the anion-exchange membrane (AEM) into the anode, while the cations (such as Na^+, K^+, Mg^{2+} and Ca^{2+}) move across the cation-exchange membrane (CEM) into the cathode chamber to maintain electro-neutrality. The MDC process can theoretically desalinate up to 90–99% of the salt content from saline water and at the same time produce more energy than the external energy required to operate the system.

In this technology, membranes play an important role in the efficiency of the MDC. A membrane could be defined as a permselective barrier separating two phases (Xu, 2005). Ion-exchange membranes (IEMs) are characterized by charged groups anchored in a polymeric matrix, which allow the pass of oppositely charged ions through the membrane, while not allowing the migration of charged ions with the same electric sign. The mobile ions inside the membrane are often called 'counter-ions', while mobile charged ions of the same sign as the fixed charges are called 'co-ions' (Krol, 1997).

The properties of ion-exchange membranes are closely related to their structures on a microscopic scale. Most hydrocarbon polymer membranes have a rather regular distribution of cross-linking and ionic groups attached to a backbone of aromatic rings. As a result, these membranes show a homogeneous structure even on a

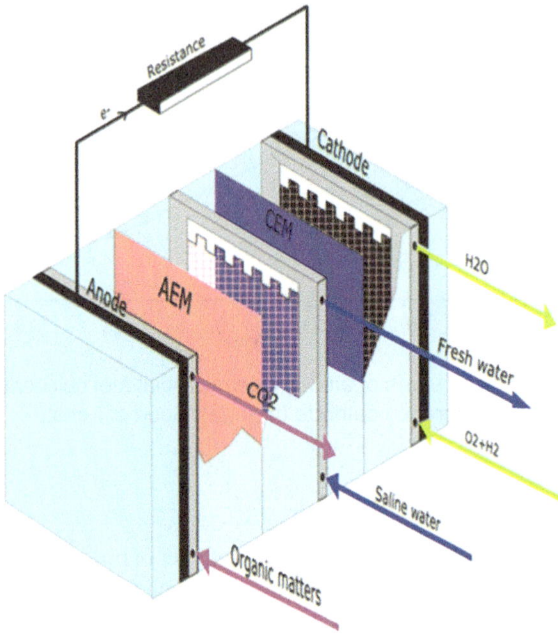

Figure 3.25 Scheme of a typical microbial desalination cell. Source: FUJIFILM.

microscopic scale. Membranes based on linear fluorocarbon polymers have a backbone of a highly crystallized polymer with ionic groups fixed on a side chain. These membranes are quite heterogeneous on a microscopic scale with distinct crystalline areas formed by the linear fluorocarbon polymer and separated by amorphous, highly hydrophilic domains formed by the side chains with the ionic groups. From experimental evidence it can be concluded that the fluorocarbon polymer membranes have a heterogeneous two-phase structure composed of crystalline polytetrafluorethylene and amorphous vinylether with fixed charges. In this sense, the amorphous region containing a cluster of fixed charges, counter-ions, and water is extremely hydrophilic and swells in water or aqueous solutions while the crystalline region acts as a cross-linkage and restricts the swelling of the membrane.

In the literature the hydrophilic clusters are pictured as spherical regions with a diameter of ca. 4 to 6 nm connected by 'bottle necks' between crystalline regions of ca. 1 nm in diameter as depicted in Figure 3.26.

The mobility of the counter-ions within the hydrophilic cluster is very high. The transport of the counter-ions from one cluster to the next, however, is restricted by the 'bottle neck'. The size of the hydrophilic cluster is a function of the swelling of the membrane and depends on the concentration of the electrolyte solution in equilibrium with the membrane (Strathmann, 2004).

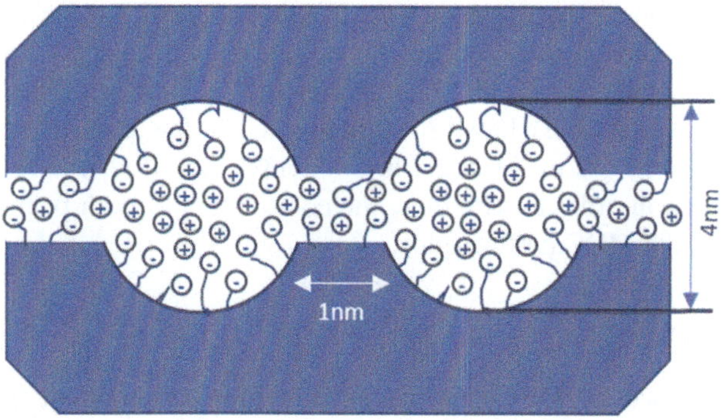

Figure 3.26 Schematic picture illustrating the cluster network of a fluorocarbon polymer based IEM. Source: adapted from Strathmann, 2004.

The environmentally friendly IEM processes are finding more and more applications in modern industries in order to demineralize concentrate and modify products. However, the main drawback of the IEM-based processes is membrane fouling, which significantly decreases the process efficiency and increases the cost (Strathmann, 2004). Fouling is the phenomenon of undesirable attachment of certain species (living organisms or non-living substances) to the surface or inside the material. The major types of IEM fouling are colloidal, organic, scaling and biofouling. Luo and coauthors (2012) reported a 27% decrease in desalination capacity of an MDC over an 8-month period. Biofouling at the AEM was the phenomenon that increased system resistance because ionic transfer was reduced, which also resulted in a decrease in energy efficiency conversion. Other problems such as scaling at the CEM were minimal. The main problem in AEMs is biofouling, while scaling is prevalent in CEMs due to the different conditions in anode and cathode chambers, respectively. The scaling on IEMs is mainly caused by the precipitation of diverse inorganic compounds such as calcium carbonate and magnesium hydroxide. In addition to biofouling, special attention should be paid to other inorganic compounds or particulate matter (non-organic matter fouling) as fouling agents are not only on the surface but also in other elements of the MDC unit (spacer, saline compartment, cathode compartment, etc.)

Mechanisms of fouling on both sides of the membranes may be different due to the contact of different solutions (in different chambers). The membrane resistance of the CEM increases more significantly because of scaling than that of the AEM with biofouling, suggesting that the CEM will require more maintenance than the AEM during MDC operation. Nevertheless, biofouling in AEMs can have a greater effect on the quality of the desalinated water. Although membrane fouling would not significantly decrease the rate of conductivity reduction within the testing period, the negative effect of fouling over a longer period of operation is expected, and thus proper control of fouling should be considered (Ping *et al.*, 2013).

Therefore, development of new IEMs, with improved antifouling behaviour among others, is needed and it is one of the most important targets. Nevertheless, there are some other factors affecting the overall resistance of the MDC unit: concentration polarization, level of mixing inside the compartments and/or Donnan potential across membranes. In this sense, overall resistance in the stack will be measured in order to determine the contribution of each phenomenon and their impact on the MDC performance.

3.2.1.1 IEMs state of the art

IEMs currently used for MDCs are manufactured for conventional electrodialysis processes and designed for good performance when electric energy is applied. However, MDCs operate at lower current; thus, the reduction of the electric resistance of the membranes can have an impact on the energy output, of great importance in some cases.

Modification of IEMs for improved properties against fouling is one the main focuses in recent literature. As mentioned previously, this phenomenon is one of the key challenges for the modern chemical, agricultural, food, pharmaceutical processing and water treatment industries. The cost of cleaning procedures and membrane replacement may vary from 20 to 30% (operational cost) for the pressure-driven membrane processes to 40–50% for the electro membrane processes. The particular ionic selectivity of IEMs leads to separation of ionic species from a neutral media allowing demineralization, concentration and modification of the resulting effluents. Moreover, IEMs are also involved in the energy conversion and storage processes. However, the formation of deposits may occur on the IEM surface and/or inside the membrane causing an increase in electrical resistance, a decrease in permselectivity, and membrane alteration. Thus, the industrial application of IEM-based processes is hampered due to the occurrence of fouling phenomena.

Different strategies have been reported to improve the resistance against fouling. They, generally, aim to change the membrane surface properties such as:

- Surface charge
- Hydrophobic/hydrophilic balance
- Roughness
- Anti-microbial layer

Modification of the AEM surface by sulfonating agents considerably improves antifouling potential. This improvement happens due to a decrease in membrane surface hydrophobicity and the introduction of negative charge density in a thin layer on the membrane surface. Moreover, using layer-by-layer modification, simultaneous improvement of antifouling potential and monovalent anion selectivity may be achieved. An increase in the cross-linking density of a CEM or the formation of a thin layer of a highly cross-linked resin on a low cross-linked

CEM results in a significant increase in permeability of monovalent over multivalent cations. The introduction of a small amount of anion-exchange groups in a CEM or the formation of a thin layer of a positively charged polyelectrolyte on the surface of a CEM also increases the membrane selectivity for monovalent ions (only on one side of the membrane surface). Tokuyama Soda has commercialized a monovalent CEM (Neosepta|CMS) prepared by forming a thin polyelectrolyte layer on the surface. The permeability ratio of sodium to calcium is more than 20 for such a membrane. Monovalent anion selective membranes have been developed mainly to prevent the permeation of sulphate which can cause scaling due to the precipitation of calcium sulphate in desalination of brackish water by electrodialysis. The selectivity of AEMs for monovalent anions can be increased by the formation of a highly cross-linked layer on the surface of the AEM. Also, the formation of a negatively charged polyelectrolyte layer on the surface of an AEM increases its selectivity for monovalent ions. The hydrophilicity of anion-exchange groups in the membrane also affects the permselectivity for specific anions. More hydrophobic anion-exchange groups have a higher permselectivity for less hydrated ions and vice versa.

Recently, a successful IEM modification by polydopamine against fouling has been reported (Vaselbehagh *et al.*, 2014). The authors also emphasized that the increase in surface negative charge and the decrease in surface hydrophobicity have a positive influence on antifouling potential, while the increase in surface roughness decreased the antifouling potential. The situation with biofouling seems to be more complex. Surface hydrophobicity and charge density are, although important factors, not effective criteria in membrane modification for biofouling control. The nature of surface charges plays the most crucial role. Membrane modification for biofouling control is carried out via two routes: an anti-adhesive approach that prevents the initial attachment of soluble microbial products and bacteria on a membrane, and an anti-microbial approach that aims to kill bacteria already attached to the membrane. An example of an anti-biofouling approach was demonstrated in a study involving membrane surface modification with biocides such as nanosilver particles (Mikhaylin & Bazinet, 2016). It is important to note that any modification to the membrane structure should have a low impact on the other IEM properties such as electrical resistance, permselectivity, and mechanical and chemical stability.

3.2.2 IEM requirements for MDC technology

IEMs developed to suit the MDC process requirements have to accomplish the following requisites: low-fouling tendency, high selectivity, increased ion-exchange capacity, reduced thickness and low ohmic resistance. In order to have a clearer idea about the targets, Table 3.12 summarizes the main features of the IEMs for application in MDCs.

Table 3.12 Required general physical-chemical properties for IEMs for MDC application.

Properties (Units)	Optimal Range	Comments
Permselectivity (%) 0.05–0.5 M KCl	>95	As high as possible. Higher than 90%
Electric resistance 2 M NaCl (Ω cm^2)	<1	As low as possible (<3)
Water content (g H_2O/g dry membrane)	0.25–0.30 <10%	As low as possible (for lower resistance) but cross-linked sufficiently for good selectivity
Thickness (μm)	100–150	The thinner the better (low resistance)
Mechanical strength	High	Free standing film
Chemical resistance	High	pH = 5–9 for AEM pH = 1–8 for CEM

Some of the most relevant specifications for both AEMs and CEMs are listed in the following sections.

3.2.2.1 Anion-exchange membranes (AEMs)

- Relevant characteristics:
  - Positively charged
  - Allow anion transport
  - Robust mechanical properties
  - Good chemical stability in alkaline environment
  - Robust device performance
  - Low water and O_2 permeability
  - Low ohmic resistance in the tested media (freshwater media, wastewater)
- Main drawbacks: substrate crossover and biofouling
- Materials mostly used: aminated polysulfone (PSU), aminated Polyvinylidene difluoride (PVDF)
- Most relevant AEMs in the market: FUJIFILM Type1/10, Ultrex AMI 7001, RALEX- AEM, Neosepta ASE.

3.2.2.2 Cation-exchange membranes (CEMs)

- Relevant characteristics:
  - Negatively charged
  - Allow cation transport

- o Low ohmic resistance in the used media
- o Robust mechanical properties
- o Good chemical stability in acidic environment
- o Robust device performance

- Main drawbacks: scaling, oxidative environment
- Materials mostly used: sulfonated polyphenylene oxide
- Most relevant CEMs in the market: FUJIFILM Type1/10, Nafion, Hyflon, Zirfon, Ultrex CMI 7000, Neosepta CSE.

3.2.3 Characterization methodology

In the frame of the MIDES project, several IEMs were developed to suite the MDC process requirements. Figure 3.27 summarizes the diverse methods employed during the project at different IEM development stages.

A brief description of the most important IEM parameters, and the specific methodologies to determine them, are described in the following sections.

3.2.3.1 Permselectivity (PS)

The permselectivity (PS) of a membrane is determined by the ratio of the flux of specific components to the total mass flux through the membrane under a given driving force. In an IEM the permselectivity is generally related to the transport of electric charges by the counter-ions to the total electrical current through the membrane. The permselectivity can be calculated from the transport or transference number of the counter- and co-ions in the membrane and the outside

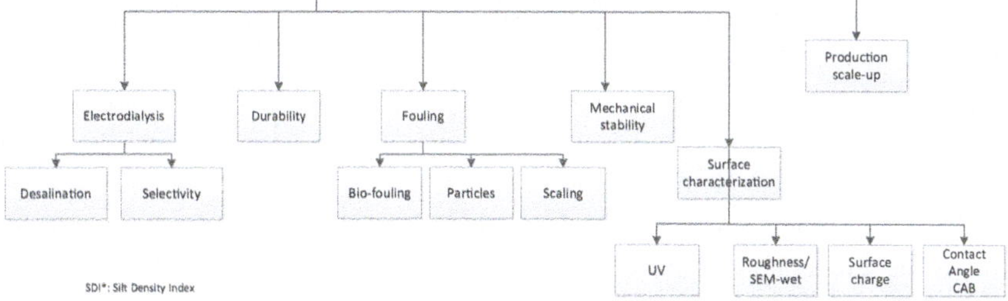

Figure 3.27 Developed evaluation methods by FUJIFILM and LEITAT. Source: FUJIFILM.

solutions. The permselectivity of a membrane is given by the following equation:

$$\Psi^m = \frac{T_{\text{cou}}^m - T_{\text{cou}}}{T_{\text{co}}} \tag{3.2}$$

where, Ψ is the permselectivity, T_{cou}^m is the transport number for counter-ions, T_{co} is the transport number for co-ions.

The transport numbers are defined by the equation:

$$T_i = \frac{z_i J_i}{\sum_i^z J_i} \tag{3.3}$$

where z is the valence, and J is the flux; the subscript i refers to cation or anion.

The PS of the membrane is a parameter that indicates the ideal degree of the membrane. Various methods to experimentally determine the permselectivity and transport number in the membrane are widely described in the literature. Generally, values ranging from 90–99% are common for IEMs used in desalination systems, with lower PS values (80%) for membranes with special properties.

PS determination is based on the measurement of potential across the selective membrane in contact with two different concentrations of ion solution (for example, 0.05–0.5 M KCl) once the equilibrium is reached.

$\Delta V = R.T/z.F.\ln(ac/ad)$, where ΔV is the theoretical potential (V), R is the gas content, T absolute temperature (K), z is valence, F is the Faraday constant ($96485\ \text{C} \cdot \text{mol}^{-1}$), and ac and ad are the chemical activities of the concentrated and diluted solutions respectively. The quotient of the measured value to the theoretical value is the PS and is expressed as a percentage.

The permselectivity was calculated using the next equation (Güler *et al.*, 2013):

$$\alpha(\%) = \frac{\Delta V_m}{\Delta V_t} \cdot 100 \tag{3.4}$$

where, α (%) is the permselectivity, ΔV_m is the measured potential difference and ΔV_t is the theoretical potential difference for a 100% permselective membrane under these experimental conditions. ΔV_t was calculated by using the Nernst equation (equation 3.5),

$$E_m = \frac{R \cdot T}{z \cdot F}(2t_i - 1)\ln\frac{a_1}{a_2} \tag{3.5}$$

where, E_m (V) is the potential difference, R ($\text{J} \cdot \text{mol}^{-1} \cdot \text{K}^{-1}$) is the gas constant, T (K) is the temperature, z is the electrovalence of the electrolyte, F ($\text{C} \cdot \text{mol}^{-1}$) is the Faraday constant, t_i is the transport number (for a 100% permselective membrane $t_i = 1$), and a_1 and a_2 are the activity coefficients of NaCl solutions (Lejarazu-Larrañaga *et al.*, 2020).

3.2.3.2 Electrical resistance (ER)

The electrical resistance (ER) of an IEM is one of the factors which determine the energy requirements of electrodialysis processes. However, in practice, most of the membranes' resistance is considerably lower than the resistance of the dilute solutions, since the ion concentration in the membrane is relatively high (Strathmann, 2004).

The electrical resistance of a membrane is determined by the ion-exchange capacity and the mobility of the ion within the membrane matrix. The ion mobility in the membrane depends strongly on the nature of the characteristics of the mobile ion species, that is, valence, size, and extent of hydration, and on the water content of the membrane. It is furthermore affected by the temperature and by the nature of the fixed ions, that is, the interaction between the mobile ions with the fixed ions.

The area resistance of an IEM can be determined by direct current (DC) measurements or by alternating current (AC) (Berezina *et al.*, 2008; Strathmann, 2004) measurements. In DC measurements the membrane is installed in a cell which consists of two chambers containing the test solution separated by the membrane being tested as indicated in Figure 3.28.

3.2.3.3 Swelling degree x,y,z

The change in IEM dimensions due to swelling are commonly expressed as a percentage change from the dry membrane in three directions (x,y,z). This factor has to be considered for a proper assembling process of IEMs into an MDC reactor.

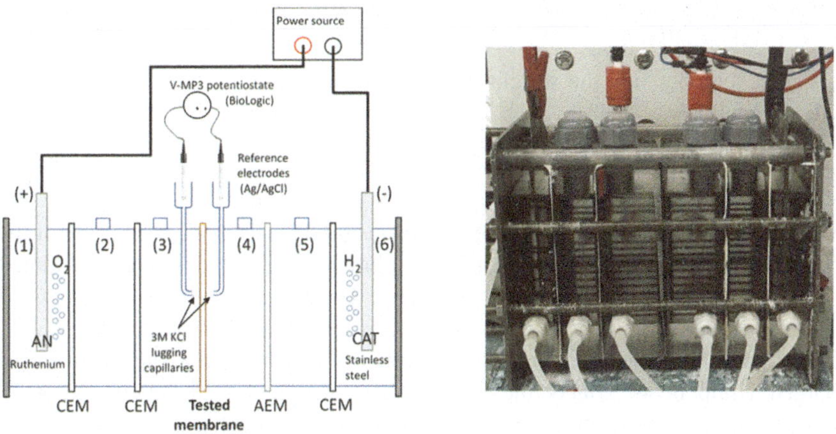

Figure 3.28 Scheme of a test cell for determining the permselectivity and electric resistance of IEMs (left) and the laboratory set-up used in LEITAT laboratories during MIDES (right). Source: Leitat.

3.2.3.4 Water uptake

The swelling of a membrane depends on a number of different parameters such as the nature of the basic membrane polymer, the nature of the ion-exchange groups and their concentration in the membrane, the counter-ions, the cross-linking density and the homogeneity of the membrane. The gel water content, th at is, swelling of a membrane, not only determines its mechanical properties and its dimensional stability, but also affects its ion permselectivity, electrical resistance, and hydraulic permeability (Strathmann, 2004).

The water content of an IEM depends not only on the properties of the membrane but also on the composition of the surrounding solution. In particular, the concentration of the solution has a significant effect on the water content of the membrane because of osmotic effects that are directly related to the chemical potential difference of the water activity between the membrane and the solution. The state of the water in the membrane can also be rather different. Part of the water is strongly bound within the hydration shell of the counter-ions and the fixed charges of the membrane, while the rest of water is more loosely bound, so-called 'free' water, to the basic polymer matrix.

The total water uptake by the membrane in equilibrium with an electrolyte solution can be determined by measuring the weight difference between a membrane in the wet and dry state. To determine the water content of a membrane, a sample is equilibrated in deionized water or any other test solution. After removing the surface water from the sample, the wet weight of the swollen membrane is determined. The sample is then dried at 30°C (17 h) until constant weight is achieved. The water content of a membrane is obtained in weight percent by:

$$WU\% = \frac{w_{wet} - w_{dry}}{w_{dry}} * 100 \quad (3.6)$$

here W_{wet} and W_{dry} are the weight of a membrane sample in the wet and the dry state, respectively.

3.2.3.5 Diffusion coefficients

When IEMs are used in electrodialysis or diffusion dialysis devices one side of the membrane is in contact with a concentrated solution and the other side with a diluted solution. In practical applications, often large concentration gradients across the membrane are obtained which results in a diffusion of low molecular weight charged and non-charged components through the membrane. The diffusion of charged components affects the current efficiency and leads eventually to a contamination of the resulting effluent or to a loss of valuable components when certain biological materials are treated (Krol, 1997). The determination of the diffusion coefficients can be carried out by standard procedures such as the time-lag method. In the time-lag experiment the membrane separates two solutions. One solution contains the test component in a certain concentration while the other solution and the membrane are free of this component. Then the flux through the membrane is measured as a function of time. In the first stage of the measurement, the so-called transition state, the flux is increasing exponentially with time. After a certain time, the flux increase with time becomes linear and the system has reached a steady state. In order to determine the extension of back-diffusion in the MDC system, the tests will be carried out in the MDC reactor in abiotic conditions using the working solutions (anolyte, catholyte and saline stream).

3.2.3.6 Fouling studies: membrane biofouling assessment

The trend to form biofilm on an IEM surface can be assessed by means of specific assays for the MDC conditions. This enables selection of the best membranes and/or the respective surface modification methods.

During MIDES, a benchmarking of biofilm formation onto modified IEMs was carried out. A microbial suspension (0.1 ml) of a model anaerobic microorganism (*Pseudomonas aeruginosa* ATCC 15442) at a known concentration (10^3 CFU) was placed onto a portion of the membrane (1 cm^2) and incubated at room temperature under anaerobic conditions for 24 hours in a wet chamber, simulating the conditions of the anodic chamber of an MDC. After that, diluent was added onto the membrane surface (0.1 ml) and incubated for a further 24 h. Membrane portions were then rinsed with distilled water twice and prepared for determining the counting of viable cells and obtaining images of the membrane surface. Thus, membrane tendency to allow biofouling was determined via two semi-quantitative methods.

In a second step, long-term experiments were carried out with the best performing membranes. Biofouling formation was promoted by exposing the membranes to the MDC anodic chamber conditions (different wastewater types) over a longer period. Microorganism adhesion and organic matter attachment was qualitatively determined during the experiments by SEM images. Finally, electrochemical characterization was carried out after fouling, to check membrane performance in terms of permselectivity and electric resistivity evolution.

The silt density index (SDI) is a widely accepted fouling index in the desalination field. FUJIFILM developed a quantitative fouling test, based on the SDI, in a dedicated system consisting of four setups running in parallel using: four membrane fouling simulators (MFSs), four peristaltic pumps and four feed tanks. In this system, a piece of membrane was placed in the MFS (4×4 cm). The adenosine triphosphate (ATP) (free and total, in relative light units) was measured for each feed tank. The feed solution circulated across the membrane surface for 17 hours at a cross velocity of 0.6 cm/s. After that, the membrane sample was rinsed gently with tap-water (tap-water has ATP close to zero). The sample was placed in 100 ml of tap-water and ultrasounds applied for 2 minutes.

The extracted ATP was measured and a reference membrane (Type 10) was included. The criterion for membrane fouling potential evaluation was the difference between total ATP and free ATP (attachment of live cells to the membrane surface in time) of the sample IEM compared to that of the reference (base membrane). The target was set at less than 30% of the reference.

3.2.4 IEMs with potential application in MDCs

Table 3.13 summarizes a selection of already commercial IEMs with potential use in MDC application, among those available in the market.

3.2.5 Manufacturing of specific IEMs for MDCs

FUJIFILM developed a cost-effective IEM production method based on an R2R dry process. Several FUJIFILM proprietary-IEM types, of both AEM and CEM, were produced during MIDES with a wide variation in the functional properties (see Table 3.14). Including certified (Type 1 & 10) and test material (Type 6 & 16) IEMs. Besides the low production cost, FUJIFILM IEMs have the advantages of being easy to transport and store, and easy to handle during the stacking and conditioning. Several of them are certified for drinking water NSF-61, including Type 1, 6 and 10.

By comparing the MDC requirements with the IEMs portfolio at FUJIFILM, Type 1 and Type 10 were listed as good candidates to be tested in the MDC. However, both types have rather high fouling tendency and special attention should be paid to the AEM. On the other hand, Type 6 shows very good antifouling performance, but it is a monovalent selective membrane, which may result in a lower desalination rate.

FUJIFILM tested several surface modifications strategies to increase their commercial IEMs' performance for MDC application. Grafting of monomers with anti-adhesive/antibacterial properties, coating of hydrophilic polymers (for surface roughness modification) with or without antibacterial nanoparticles, and treatment with surfactants, are some of the strategies that were evaluated for membrane modification.

Many different samples were produced in the R2R test coating machine during MIDES, all one-metre length and 15 cm width. Of the produced levels, several samples showed good performance in the antifouling tests. Two lead candidates were initially identified: (1) coating based on charged monomer (2) 1:1 mixture of non/charged monomer. Both candidates were scaled up to production scale.

3.2.6 Membrane testing at lab-scale under operative conditions

FUJIFILM tested real wastewaters to evaluate their fouling potential and to investigate the performance of the developed antifouling membrane versus the base membranes Type 1, 6 and 10. An industrial brewery wastewater (influent)

Table 3.13 Properties and prices of the IEMs commercially available.

Membrane (Company)	Type	PS (%)*	ER (Ohm cm^2)†	Thickness (mm)‡
CMX-SB (Neosepta)	Homogeneous	98	1.8–3.8	0.17
AMX-SB (Neosepta)	Homogeneous	98	2.0–3.5	0.14
CMI–7000S (Membrane International)	Heterogeneous	94	<30	0.45
AMI–7001S (Membrane International)	Heterogeneous	90	<40	0.45
61AZ-L386 (Ionics)	Heterogeneous	–	6	0.5
204SX-ZL386 (Ionics)	Heterogeneous	–	7	0.5
CM(H)-PES (Ralex)	Heterogeneous	>90	<7.5	<0.45
AM(H)-PES (Ralex)	Heterogeneous	>90	<7.5	<0.45
Fumasep-FKS (Fumatech)	Homogeneous	>96	<8	0.11–0.13
FAS–130-PET (Fumatech)	Homogeneous	94.7	3.1	0.125–0.135
Type 1 CEM (FUJIFILM)	Homogeneous	92–94	1.2–1.7	0.15
Type 1 AEM (FUJIFILM)	Homogeneous	90–95	0.7–0.9	0.15
Type 10-AEM (FUJIFILM)	Homogeneous	1.2–1.5	0.15	1–13
Type 10-CEM (FUJIFILM)	Homogeneous	1.2–1.5	0.15	1–13

*PS test conditions: Neosepta (N/A), Membrane International (0.5/0.1 M KCl/kg), Ionics (N/A), Ralex (0.5/0.1 M KCl), Fujifilm (0.5 M KCl).
†ER test conditions: Neosepta (0.5 N solution, 25°C), Membrane International (0.5 M NaCl), Ionics (N/A), Ralex (0.5 M NaCl), Fujifilm (2 M NaCl).
‡Reference condition (21°C, 50% RH).

Table 3.14 IEMs provided by FUJIFILM for MDC purposes in MIDES project.

Membrane	PS 0.05–0.5 M KCl (%)	ER 2 M NaCl (Ohm · cm^2)	Thickness 21°C/50 RH (mm)	pH Range
Type 1-AEM	90–95	0.7–0.9	0.15	2–10
Type 1-CEM	92–94	1.2–1.7	0.15	4–12
Type 10-AEM	97	1.2–1.5	0.15	1–13
Type 10-CEM	98	1.2–1.5	0.15	1–13
Type 6-AEM (test material)	94	0.97–1	0.15	2–10
Type 16-AEM (test material)	97	1.2–1.5	0.15	2–12

and the effluent of the anaerobic reactor treating it (effluent) were subjected to static SDI and dynamic (bio)fouling tests. The feed and effluent streams of the anaerobic bioreactor treating brewery (industrial) wastewater were characterized prior to the tests (Table 3.15).

Three types of wastewater were tested using FUJIFILM quantitative protocol: influent and effluent of the anaerobic bioreactor treating brewery wastewater and filtrated industrial membrane bioreactor (MBR) effluent containing a mixture of different live microorganisms. Two Type 10 membrane modified candidates were evaluated: Candidate 1 and Candidate 2. Table 3.16 summarizes the results from the dynamic-biofouling test.

Results did not indicate significant bacterial adhesion reduction of the modified membranes compared to reference membranes (Type 10), as shown in Table 3.16. The effects of the anti-fouling layer could not be effectively quantified due to the low fouling level of the base membrane Type 10.

The fouling potential of the effluent was further investigated using three different membranes: Type 1, 6 and 10. Results (see Table 3.17) confirmed that the effluent had a high biofouling potential on Type 1. However, type 6 and 10 showed a very low tendency for biofouling.

LEITAT conducted biofouling tests with real wastewaters (filtered and non-filtered molasses and anaerobic sludge from a wastewater treatment plant) to

Table 3.15 Feed characterization for fouling tests.

Solution	pH	Conductivity (mS/cm)	Turbidity (NTU)	COD (mgO_2/l)	PgATP/ml
Influent	8.25	3	–	1150	996
Effluent	7.67	3	48.5	565	23,469

Table 3.16 Summary of the dynamic-biofouling tests for real wastewaters, before and after treatment in an MBR, using modified type 10 FUJIFILM membranes.

Membrane	Influent		Effluent	
	PgATP/ ml · cm²	% Respect Reference	PgATP/ ml · cm²	% Respect Reference
Type 10 raw	1.6	–	-1.1	–
Candidate 1	–	58.1	–	159.5
Candidate 2	–	101.3	–	90.9

assess the biofouling due to bacterial attachment under desalination operating conditions. Tests were undertaken in a laboratory set-up replicating the operating conditions of an operating MDC. For such a purpose a bicameral system was specifically designed mimicking the anodic and saline chamber of the MDC. The tested AEMs were placed between the two chambers, with a contact area of 100 cm^2. Pictures (Figure 3.29) and SEM images (Figure 3.30) enabled qualitative determination of the fouling. AEMs Type 6 and Type 10 showed the best performance in terms of biofouling prevention, see Table 3.18. Visual observation of membrane fouling could be further confirmed by SEM images, see Figure 3.30, which confirmed bacterial adhesion on the surface during the test.

The proper IEM selection for MDC application is a critical step. MIDES provided several tools to select the most convenient membrane for each case study. Fouling tendency is likely to depend both on membrane characteristics and wastewater/effluent composition. Thus, a tailored selection process is mandatory to select the most appropriate membrane.

3.2.7 Applications of IEMs for MDC application

FUJIFILM developed IEMs were submitted to a benchmarking test. It consisted of determining the desalination performance of a synthetic brackish water in an air-cathode MDC. Figure 3.31 shows that after 72 h operation, the Type 10 IEM

Table 3.17 Summary of the dynamic-biofouling tests for real wastewaters using different FUJIFILM AEMs exposed to the effluent of an MBR.

Membrane	Effluent PgATP/ml · cm²
Type 1	31
Type 6	<1
Type 10	0

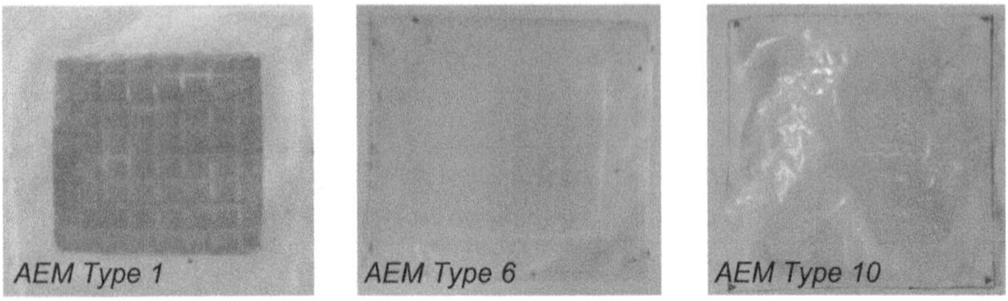

Figure 3.29 Images of the different AEMs after biofouling tests. Source: Leitat.

and its lead candidates (modifications #16 and #2) reached slightly better desalination values than reference IEMs (Membrane International and Neosepta). Thus, reaching the MIDES project goal of achieving a final salt concentration after MDC of 5 g/l, the prerequisite to reduce the posterior reverse osmosis energy consumption.

The measurement of the associated water loss (Figure 3.32) enables the recalculation of the real ions removal rates, resulting in the goal completion in all cases. Regarding the water loss, Type 10 and its modifications achieve good results in the best range of reference membranes. The best performing membranes, despite the water trespassing, were Membrane International IEMs closely followed by Type 10 IEMs, which achieved the second-best desalination rate accounting for 2.2 g-NaCl/($m^2 \cdot h$) with a 58.3% salt removal (Figure 3.32). This is a slightly better desalination performance compared with the anti-biofouling modifications and with Neosepta membranes.

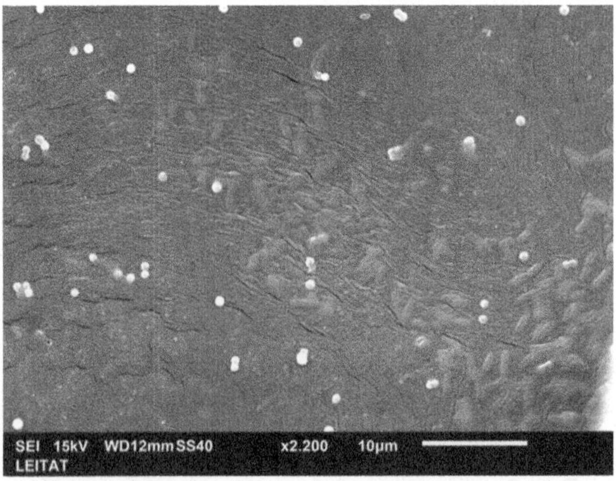

Figure 3.30 Images of biofilm and microorganisms attached to the AEM type 1 surface after biofouling assay. Source: Leitat.

Table 3.18 Summary of the biofouling tests under MDC operating conditions for different real wastewaters using different FUJIFILM AEMs.

Wastewater	AEM Type 1	AEM Type 6	AEM Type 10
Filtered molasses	Fouled	Non-fouled	Non-fouled
Non-filtered molasses	Fouled	Non-fouled	Non-fouled
Anaerobic sludge	Fouled	Fouled	Fouled

3.3 SUMMARY

Electrode development has enabled a deeper understanding of the use of different carbon-based materials for MDCs. On the one hand, a relevant amount of anode materials, with different composition and structure, have been successfully assessed for electroactive biofilm attachment. On the other hand, different cathode materials have been developed for liquid and air-cathode MDC architectures. Best operational performance has been achieved with materials for the liquid catholyte approach. Nevertheless, air-cathode materials, based on metal catalysts, proved their capacity, in terms of ORR, to be employed to drive brackish water desalination. A first attempt at upscaling air-cathodes for MDCs was achieved during MIDES (see also Chapter 2).

Ion-exchange membranes for MDC application have been the other main pillar for MDC improvement. Efforts during the MIDES project have enabled knowledge to be gained on different strategies to produce membranes with specific properties for fouling prevention. The different membrane modifications have been assessed employing both quantitative and qualitative tools. Thus,

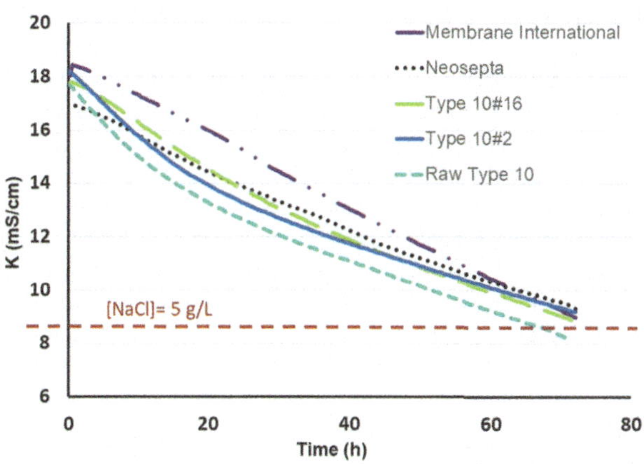

Figure 3.31 MDC saline chamber conductivity evolution over time for each IEM tested. Source: Leitat.

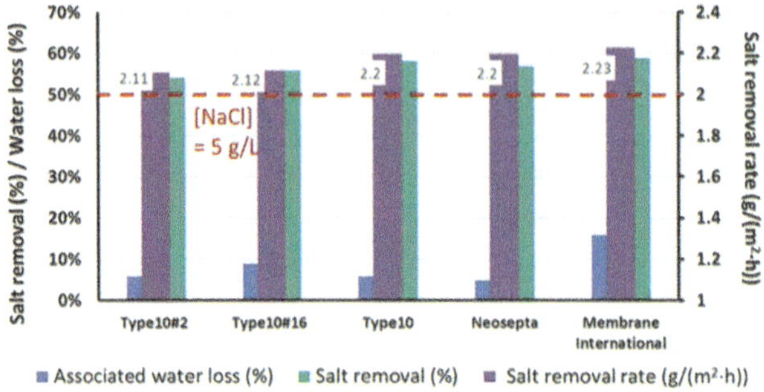

Figure 3.32 MDC desalination parameters for the tested IEMs. Source: Leitat.

allowing selection of the most promising candidates for pilot-scale production and testing in desalination trials in MDCs.

The selection of electrodes and membranes for the MDC piloting was based on the outcomes of the present work. Carbon felt KFD2.5 (SGL), was employed for both anode and cathode. Whereas NS-type isostatic graphite (SGL) was employed as the current collector for both cases. As for membranes, FUJIFILM Type 10 was selected, specifically candidate 1 as the AEM.

In conclusion, the cooperation among SGL, FUJIFILM, LEITAT and IMDEA during MIDES permitted optimization of key elements for optimal MDC performance. The gained knowledge is not only relevant for the microbial desalination cell research field but also for other microbial electrochemical technologies.

REFERENCES

Berezina N. P., Kononenko N. A., Dyomina O. A. and Gnusin N. P. (2008). Characterization of ion-exchange membrane materials: properties vs structure. *Advances in Colloid and Interface Science*, **139**(1–2), 3–28, https://doi.org/10.1016/j.cis.2008.01.002

Bhattacharjee U. and Pandey L. M. (2020). Novel nanoengineered materials-based catalysts for Various bioelectrochemical systems [chapter]. *ACS Symposium Series*, **1342**, 45–71, https://doi.org/10.1021/bk-2020-1342.ch003

Borràs E., Aliaguilla M., Bossa N., Martinez-Crespiera S., Huidobro L., Schweiss R., Schwenke A. and Bosch-Jimenez P. (2021). Nanomaterials-based air-cathodes use in Microbial Desalination Cells for drinking water production: synthesis, performance and release assessment. *Journal of Environmental Chemical Engineering*. In press.

Bosch-Jimenez P., Martinez-Crespiera S., Amantia D., Pirriera M. D., Forns I., Shechter R. and Borràs E. (2017). Non-precious metal doped carbon nanofiber air-cathode for microbial fuel cells application: oxygen reduction reaction characterization and long-term validation. *Electrochimica Acta*, **228**, 380–388, https://doi.org/10.1016/j.electacta.2016.12.175

Cai T., Meng L., Chen G., Xi Y., Jiang N., Song J., Zheng S., Liu Y., Zhen G. and Huang M. (2020). Application of advanced anodes in microbial fuel cells for power generation: a review. *Chemosphere*, **248**, 125985–126000, https://doi.org/10.1016/j.chemosphere.2020.125985

Cao X., Huang X., Liang P., Xiao K., Zhou Y., Zhang X. and Logan B. E. (2009). A new method for water desalination using microbial desalination cells. *Environmental Science & Technology*, **43**(18), 7148–7152, https://doi.org/10.1021/es901950j

Gergin I., Ismar E. and Sarac A. S. (2017). Oxidative stabilization of polyacrylonitrile nanofibers and carbon nanofibers containing graphene oxide (GO): A spectroscopic and electrochemical study. *Beilstein Journal of Nanotechnology*, **8**, 1616–1628. https://doi.org/10.3762/bjnano.8.161 (Accessed: November 2020).

Ghasemi M., Shahgaldi S., Ismail M., Kim B. H., Yaakob Z. and Wan Daud W. R. (2011). Activated carbon nanofibers as an alternative cathode catalyst to platinum in a two-chamber microbial fuel cell. *International Journal of Hydrogen Energy*, **36**(21), 13746–13752, https://doi.org/10.1016/j.ijhydene.2011.07.118

Ghimire P. C., Schweiss R., Scherer G. G., Lim T. M., Wai N., Bhattarai A. and Yan Q. (2019). Optimization of thermal oxidation of electrodes for the performance enhancement in all-vanadium redox flow battery. *Carbon*, **155**, 176–185, https://doi.org/10.1016/j.carbon.2019.08.068

Güler E., Elizen R., Vermaas D. A., Saakes M. and Nijmeijer K. (2013). Performance-determining membrane properties in reverse electrodialysis. *Journal of Membrane Science*, **446**, 266–276, https://doi.org/10.1016/j.memsci.2013.06.045

Gyenge E. L. and Drillet J.-F. (2011). The electrochemical behavior and catalytic activity for oxygen reduction of mnO_2/C– toray gas diffusion electrodes. *Journal of The Electrochemical Society*, **159**(2), F23–F34, https://doi.org/10.1149/2.061202jes

Huong Le T. X., Bechelany M. and Cretin M. (2017). Carbon felt based-electrodes for energy and environmental applications: a review. *Carbon*, **122**, 564–591, https://doi.org/10.1016/j.carbon.2017.06.078

Krol J. J. (1997). Monopolar and Bipolar Ion Exchange Membranes – Mass Transport Limitations. 164.

Lejarazu-Larrañaga A., Molina S., Ortiz J. M., Riccardelli G. and García-Calvo E. (2020). Influence of acid/base activation treatment in the performance of recycled electromembrane for fresh water production by electrodialysis. *Chemosphere*, **248**, 126027–126036, https://doi.org/10.1016/j.chemosphere.2020.126027

Lopez-Atalaya M., Codina G., Perez J. R., Vazquez J. L. and Aldaz A. (1992). Optimization studies on a Fe/Cr redox flow battery. *Journal of Power Sources*, **39**, 147–154. https://doi.org/10.1016/0378-7753(92)80133-V

Luo H., Xu P., Jenkins P. E. and Ren Z. (2012). Ionic composition and transport mechanisms in microbial desalination cells. *Journal of Membrane Science*, **409–410**, 16–23, https://doi.org/10.1016/j.memsci.2012.02.059

Majidi M. R., Shahbazi Farahani F., Hosseini M. and Ahadzadeh I. (2019). Low-cost nanowired α-mnO2/C as an ORR catalyst in air-cathode microbial fuel cell. *Bioelectrochemistry*, **125**, 38–45, https://doi.org/10.1016/j.bioelechem.2018.09.004

Mano N. (2020). Recent advances in high surface area electrodes for bioelectrochemical applications. *Current Opinion in Electrochemistry*, **19**, 8–13, https://doi.org/10.1016/j.coelec.2019.09.003

Merino-Jimenez I., Santoro C., Rojas-Carbonell S., Greenman J., Ieropoulos I. and Atanassov P. (2016). Carbon-based air-breathing cathodes for microbial fuel cells. *Catalysts*, **6**, 1–13. https://doi.org/10.3390/catal6090127

Mikhaylin S. and Bazinet L. (2016). Fouling on ion-exchange membranes: classification, characterization and strategies of prevention and control. *Advances in Colloid and Interface Science*, **229**, 34–56, https://doi.org/10.1016/j.cis.2015.12.006

Minke C., Kunz U. and Turek T. (2017). Carbon felt and carbon fiber – a techno-economic assessment of felt electrodes for redox flow battery applications. *Journal of Power Sources*, **342**, 116–124, https://doi.org/10.1016/j.jpowsour.2016.12.039

Oh S., Min B. and Logan B. E. (2004). Cathode performance as a factor in electricity generation in microbial fuel cells. *Environmental Science & Technology*, **38**, 4900–4904. https://doi.org/10.1021/es049422p

Ping Q., Cohen B., Dosoretz C. and He Z. (2013). Long-term investigation of fouling of cation and anion exchange membranes in microbial desalination cells. *Desalination*, **325**, 48–55, https://doi.org/10.1016/j.desal.2013.06.025

Ramírez-Moreno M., Rodenas P., Aliaguilla M. and Bosch-Jimenez P. (2019). Comparative performance of microbial desalination cells using air diffusion and liquid cathode reactions: study of the salt removal and desalination efficiency. *Frontiers in Energy Research*, **7**, 1–12, https://doi.org/10.3389/fenrg.2019.00135

Roche I. and Scott K. (2009). Carbon-supported manganese oxide nanoparticles as electrocatalysts for oxygen reduction reaction (orr) in neutral solution. *Journal of Applied Electrochemistry*, **39**(2), 197–204, https://doi.org/10.1007/s10800-008-9653-9

Santoro C., Stadlhofer A., Hacker V., Squadrito G., Schröder U. and Li B. (2013). Activated carbon nanofibers (ACNF) as cathode for single chamber microbial fuel cells (SCMFCs). *Journal of Power Sources*, **243**, 499–507, https://doi.org/10.1016/j.jpowsour.2013.06.061

Santoro C., Talarposhti M. R., Kodali M., Gokhale R., Serov A., Merino-Jimenez I., Ieropoulos I. and Atanassov P. (2017). Microbial desalination cells with efficient platinum-group-metal-free cathode catalysts. *ChemElectroChem*, **4**, 3322–3330. https://doi.org/10.1002/celc.201700626

Schweiss R. (2020). Validation of 1D porous electrode theory using steady-state measurements of flooded electrodes at variable electrolyte compositions. *Chemical Engineering Science*, **226**, 1–7, https://doi.org/10.1016/j.ces.2020.115841

Strathmann H. (2004). Ion-exchange membrane separation processes. *Membrane Science and Technology Series*, **9**, https://doi.org/10.1016/S0927-5193(13)70006-8

Vaselbehagh M., Karkhanechi H., Mulyati S., Takagi R. and Matsuyama H. (2014). Improved antifouling of anion-exchange membrane by polydopamine coating in electrodialysis process. *Desalination*, **332**(1), 126–133, https://doi.org/10.1016/j.desal.2013.10.031

Xu T. (2005). Ion exchange membranes: state of their development and perspective. *Journal of Membrane Science*, **263**(1–2), 1–29, https://doi.org/10.1016/j.memsci.2005.05.002

Yuan H., Hou Y., Abu-Reesh I. M., Chen J. and He Z. (2016). Oxygen reduction reaction catalysts used in microbial fuel cells for energy-efficient wastewater treatment: a review. *Materials Horizons*, **3**(5), 382–401, https://doi.org/10.1039/c6mh00093b

Zhou M., Chi M., Luo J., He H. and Jin T. (2011). An overview of electrode materials in microbial fuel cells. *Journal of Power Sources*, **196**, 4427–4435, https://doi.org/10.1016/j.jpowsour.2011.01.012

Chapter 4

Design of the MIDES plant

Juan Arévalo, Patricia Zamora, Vicente F. Mena, Naiara Hernández-Ibáñez, Victor Monsalvo-Garcia and Frank Rogalla

Aqualia, Av. del Camino de Santiago, 40, edificio 3, 4ª planta, 28050, Madrid, Spain

ABSTRACT

This chapter presents the full design of two microbial desalination cells (MDCs) at pilot-plant scale from the MIDES project. The final MDC pilot unit design was based on the knowledge gained through up scaling of the MDC from lab- to pre-pilot scale. The MDC pilot plant consists of one stack of 15 MDC pilot units with 0.4 m^2 electrode area. This chapter also presents the piping and instrumentation diagram (P&ID) and layout of the MDC pilot plant. The MIDES pilot plants are comprised of an MDC pilot plant housed in a 40-ft container with the rest of the peripheral elements. Finally, this chapter presents the improvement made from the first to the second MDC stack in terms of stability and the chemical compatibility of the end plates.

Keywords: Desalination, microbial desalination cell, pilot plants, water post-treatment, water pre-treatment

doi: 10.2166/9781789062120_0093

4.1 OVERVIEW OF GUIDELINES FOR THE DESIGN OF THE MIDES TECHNOLOGY

One of the project's main objectives is to develop the first MDC pilot plant and thereby achieve an improvement in the production of desalinated drinking water, minimizing the energy cost. During the development of the MIDES project, a careful scaling process was carried out to gain knowledge for the further design, construction, and operation of MDC pilot plants. As detailed in Figure 4.1, the roadmap of the lab-MDC upscaling (100 cm^2 electrode area) goes through the assembly of a pre-pilot modular MDC comprising 10–20 units of 600 cm^2 electrode area (see Chapter 2), towards the development of the world's largest MDC technology (15 units of 4000 cm^2).

As is indicated in Chapter 2, the starting point was the development and study of a lab-scale MDC of 100 cm^2 (10 × 10 cm). A series of desalination assays were performed mimicking the operational conditions established for the MDC pilot in different scenarios. The studies performed showed higher nominal desalination rates than those reported to date (Zhang & He, 2015).

After acquiring knowledge through the lab-scale MDC studies, the next step was the development of a pre-pilot MDC. In this step, all the project partners worked together in its design, construction, and operation. First a unit cell pre-pilot scale MDC was constructed, following the design of the lab-scale MDC, with an area of 600 cm^2 (21 × 29 cm) (see Figure 4.1(b)). Different materials and configurations were tested in order to obtain the maximum information possible to scale up the technology effectively. Once all the components were optimized,

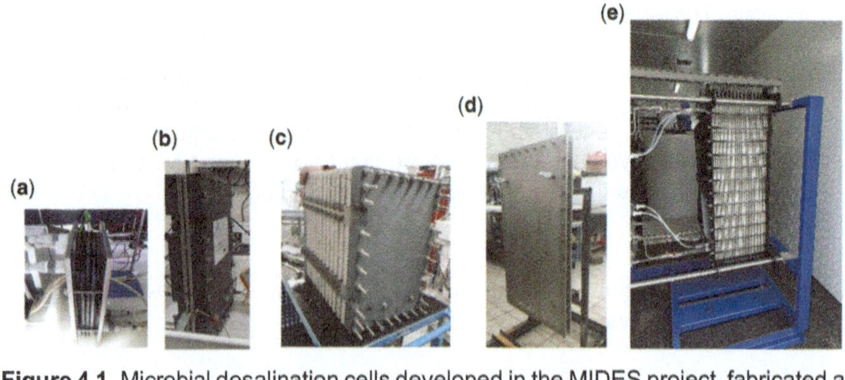

Figure 4.1 Microbial desalination cells developed in the MIDES project, fabricated at the following scales: (a) MDC laboratory area 100 cm^2; (b) pre-pilot MDC unit, area 600 cm^2; (c) pre-pilot MDC stack, area 600 cm^2 × 10–20 units (6000–12,000 cm^2, modular system); (d) MDC pilot unit, area 4000 cm^2 and (e) MDC pilot stack, area 4000 cm^2 × 15. Source: Author's own work/property.

Table 4.1 Materials up-scaled for the MDC pilot plant construction.

	Membranes		**Electrodes**		**Plastic materials**
Element	**Anionic Exchange Membrane (AEM)**	**Cationic Exchange Membrane (CEM)**	**Anodic and Cathodic Carbon Felt**	**Anodic and Cathodic Current Collector**	**MDC Back Plates**
Partner	Fujifilm	Fujifilm	SGL	SGL	Mikrolin
Type/identification	Type 10	Type 10	KFD 2.5	Isostatic graphite	Plastic waste mixture flat sheets
Size (m)	1.05 × 0.53	1.05 × 0.53	0.44 × 0.92	0.50 × 1.25	1.10 × 0.50
Thickness (mm)	0.15	0.16	2.50	4.00	10.00
Fabrication technique/procedure	Coating	Coating	Roll to roll needling and carbonization process	Classification milling of calcined petroleum coke, mixing of coke and coal tar pitch under heat, cooling, milling/sieving, isostatic pressing, baking, graphitization, slicing, vacuum impregnation with liquid thermoset resins, curing, machining	Shredding/grinding, separation/cleaning, dry mixing, regranulation/blending, extrusion/compression moulding

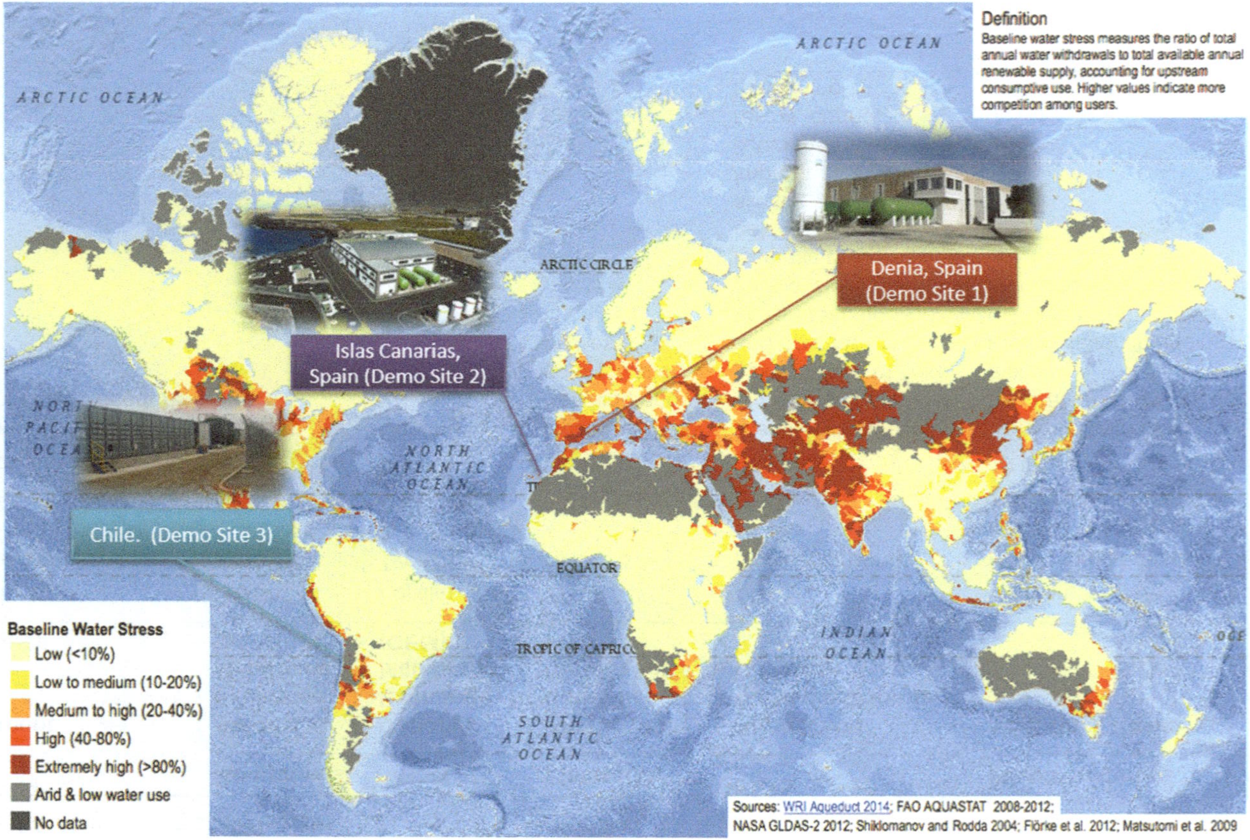

Figure 4.2 Demo sites chosen for the validation of the two MDC pilot plants. Demo Site 1: BWDP Racons (Denia, Alicante) and Demo Site 2: SWDP Oeste (Guía de Isora, Tenerife).

the same desalination assays performed with the lab-scale MDC were carried out, working with the different streams which will be present in full-scale operation (brackish water, seawater, synthetic saline water, urban wastewater, industrial wastewater, saline wastewater). The results showed a good scalability of the system. Following this achievement a 10 unit cell pre-pilot MDC stack was constructed (Figure 4.1(c)). Once each of the components required for the correct assembly of the MDC stack was optimized, the MDC was tested by means of the same desalination assays performed previously. Results once again showed the good scalability of the system. This information has been key for the operation and optimization of the final MDC pilot.

Once the development of both the single unit and the stack pre-pilot MDC was successful, the design of the pilot MDC was completed. In this case, the MDC consisted of a set of 15 cells each with an area of 4000 cm^2, with the same approach as in the pre-pilot stack, monopolar connections, and working each cell in parallel. Components of the MDC stack such as the electrodes, membranes, and plastic materials were upscaled at a sufficiently large scale for the construction of the MDC pilot stack (Table 4.1). The membranes, electrodes, and plastic components upscaling process is detailed in Chapter 3.

This technology will be validated at the pilot scale in two demo sites: Demo Site 1, the brackish water desalination plant (BWDP) Racons and Demo Site 2, seawater desalination plant (SWDP) Fonsalia Guía de Isora, Spain, for brackish water and seawater desalination, respectively (Figure 4.2). Two pilot plants have been constructed and operated under real environments in desalination plants operated by Aqualia. The first MDC pilot plant is located in the research centre desalination technologies of Aqualia in BWDP Racons in Denia (Spain). At Demo Site 1 the MIDES technology will be validated by studying the desalination of brackish water from the Racons river. The other demo site chosen was SWDP Oeste, in Guía de Isora (Tenerife). In this case, the partial desalination of seawater using the MDC, for its subsequent post-treatment using a reverse osmosis system, will be studied.

4.2 MDC DESCRIPTION AND DESIGN

The microbial desalination cell is the central element in the MIDES treatment scheme, where both water streams are simultaneously treated by electroactive bacteria. The MIDES pilot plants are comprised of an MDC pilot plant housed in a 40-ft container which consists of:

- One MDC pilot stack comprising 15 cells of 0.4 m^2 electrode area.
- Anolyte tank, volume 1 m^3.
- Catholyte ($FeCl_3$) tank, volume 1 m^3.
- Brackish water tank, volume 350 L.
- Anolyte pump, max flowrate 500 L/h.

- Catholyte pump, max flow rate 500 L/h.
- Brackish water pump, max flow rate 500 L/h.

The piping and instrumentation diagram (P&ID) of the MDC pilot stack is shown in Figures 4.3 and 4.4 presents the layout of the MDC with the distribution of the plant inside the container.

The MDC can work as a stand-alone desalination device or as a pre-desalination prior to reverse osmosis (RO). The mode of operation will be batchwise. The saline stream, the wastewater and the catholyte ($FeCl_3$) will be recirculated through the

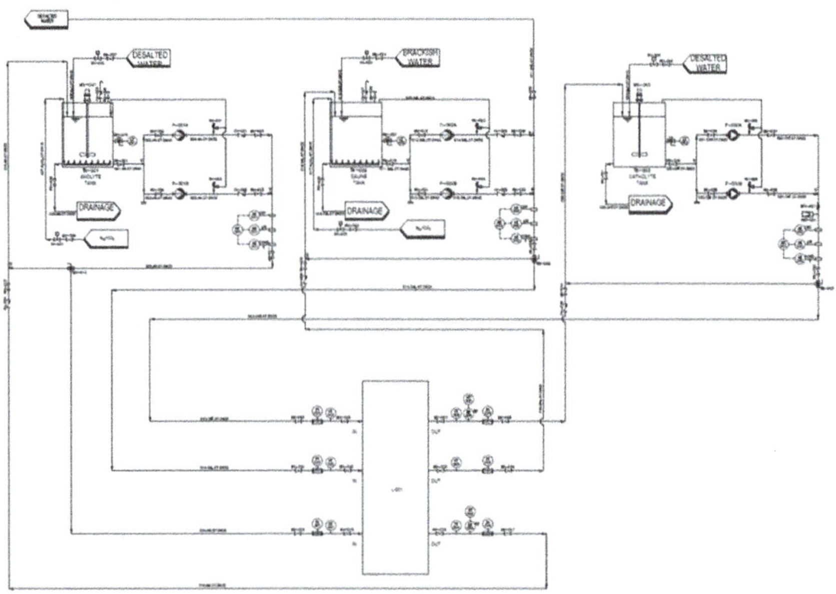

Figure 4.3 P&ID of MDC pilot stack, Demo Site 1 Denia. Source: Engineering department of Aqualia.

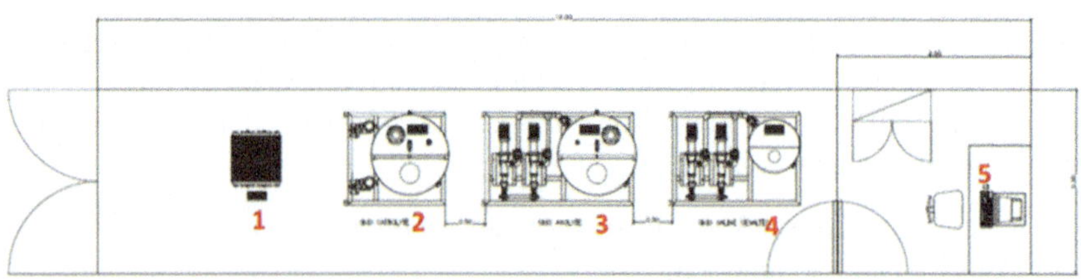

Figure 4.4 Layout of MDC pilot stack container. (1) MDC pilot stack, (2) Skid catholyte, (3) Skid anolyte, (4) Skid saline, and (5) Workplace. Source: Engineering department of Aqualia.

MDC stack until the saline stream reaches the target conductivity for partial or complete desalination. Once the target conductivity is achieved, the desalted water will be stored in a tank for subsequent RO and/or post-treatments.

The MDC pilot plant was designed taking into consideration the information obtained at previous scales and revised by a specialist. The MDC pilot consists of a stack of 15 cells with an area of 4000 cm^2 each, with an approach similar to that of the pre-pilot MDC, monopolar electrical connection and working each cell in parallel. Based on the results obtained previously, a series of changes were proposed for the construction of the pilot MDC, necessary to solve the issues that arose with the increase in the area and the number of cells.

Firstly, the MDC pilot design was initially based on a concept of stacking gasket-like layers with gasket sealing. With the intention to improve the isolation between compartments, the gasket configuration was replaced by an O-ring configuration. A specific design that was conceptually developed for a multi-chambers stack was proposed, which is based on the key concept of having

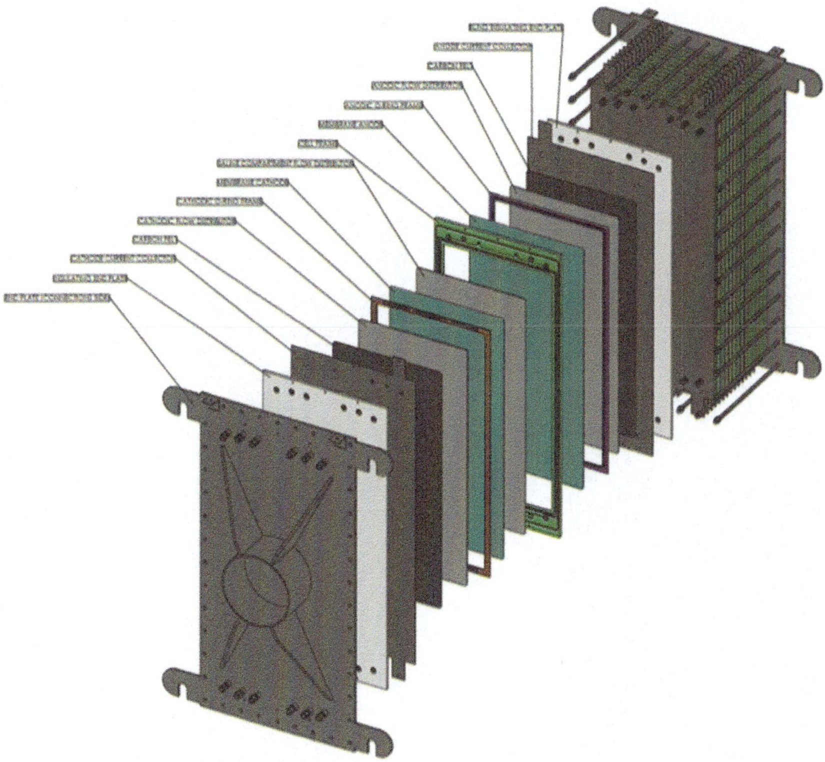

Figure 4.5 Overall MDC pilot design for Demo Site 1. Layout of the MDC stack assembly. Source: Hysytech S.r.l.

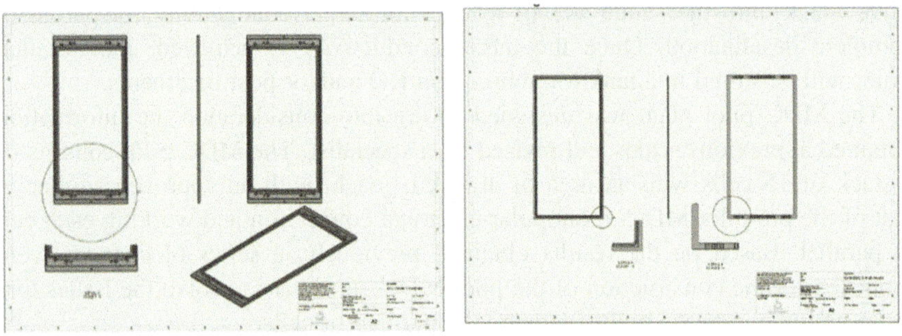

Figure 4.6 Main frame detail of MDC pilot unit and O-ring frame detail of MDC pilot unit. Source: Hysytech S.r.l.

enough rigid polymeric layers to be sealed by means of O-rings. The coupling of rigid materials with soft and easily compressible O-ring allows application of a defined strength to determine a significant deformation of the overall stack size until the rigid components touch each other. At that point the sealing is guaranteed by

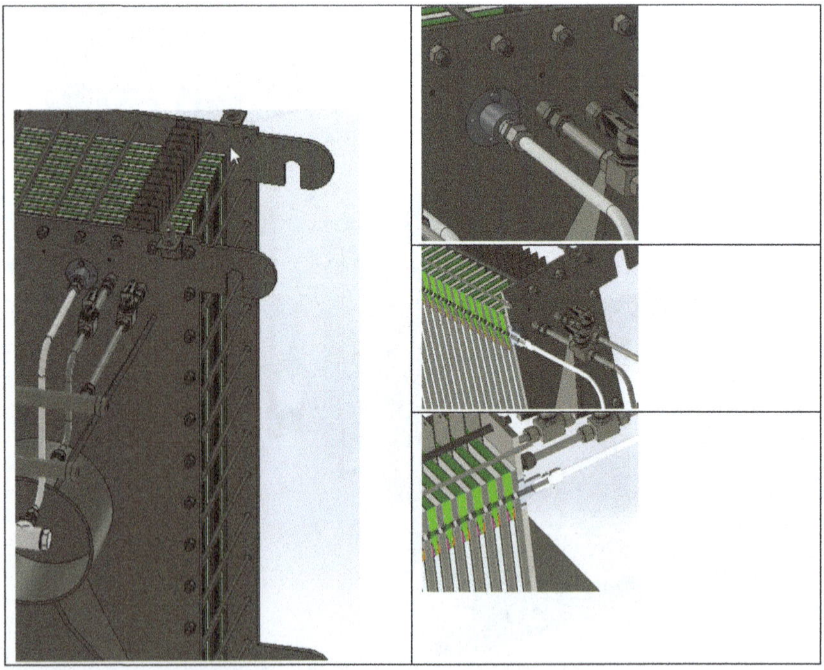

Figure 4.7 Redesign of one of the four connections of the cathodic line for the second MDC. Source: Hysytech S.r.l.

properly compressed O-rings in suitable seats and, on the other hand, the rigid components preserve the volume of distribution channels and volume occupied by inner component (electrodes, flow distributor, current collector, and membranes).

Secondly, the overall strategy is to operate the MDC unit in a high recirculation regime. This is beneficial in terms of relaxing the performance required by each cell. The main requirement in this case is to allow a low enough pressure drop to minimize the pressure difference between the inlets and the outlets of each of the chambers (anodic, cathodic and saline).

The main modifications made to the original design are summarized here:

(a) Replacement of silicon gaskets by O-rings.
(b) Flow distributor in anode, saline and cathode compartments (the same for the three chambers).
(c) Frontal feed for anolyte, saline and catholyte and two inlets per stream to improve flow distribution.
(d) Thickness of the saline compartment: 2.5 mm.

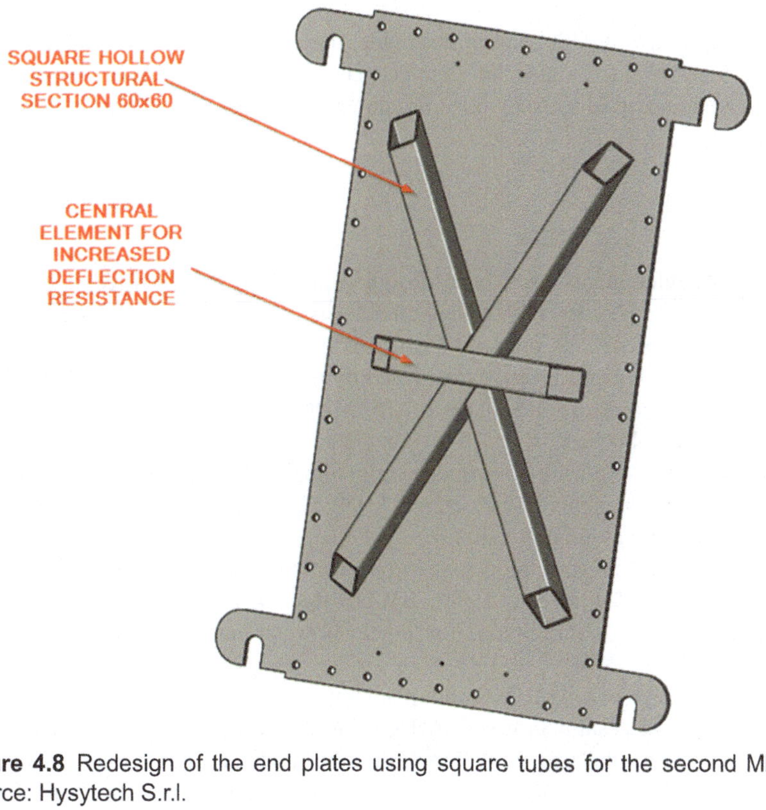

Figure 4.8 Redesign of the end plates using square tubes for the second MDC. Source: Hysytech S.r.l.

(e) Thickness of the anolyte and catholyte compartments: 5 mm (2.5 mm carbon felt + 2.5 mm flow distributor).
(f) End plates and compartments made of polypropylene (PP).
(g) Carbon felt on anode and cathode chamber: KFD 2.5 SGL.
(h) Current collector: NS-type isostatic graphite SGL.

The overall MDC pilot unit designed is shown in Figure 4.5. In Figure 4.6 the frame detail of the MDC pilot unit and O-ring frame detail of the MDC pilot unit are shown.

In order to improve the chemical compatibility of the MDC stack with the process fluids (in particular ferric chloride), piping, collectors and the fittings of the cathodic line were redesigned. PVC was selected for the fabrication of the new components because of its chemical compatibility with the fluids inside the cell and its commercial availability. Since the other parts of the stack are not in contact with ferric chloride, no changes were necessary. Figure 4.7 shows the modifications performed in the second MDC design regarding the chemical compatibility of the catholyte line. The figure shows the redesign of the inlet and outlet piping and connections of the cathodic line to the MDC stack end plate, all of them designed in PVC.

In order to avoid deflection in the frame, the end plates were redesigned using square tubes welded onto the flanges. The preliminary software simulations suggested good results using 60×60 mm tubes in the configuration shown in Figure 4.8.

4.3 CONCLUSIONS

Two MDC pilot plants have been designed based on the knowledge gained in previous steps. A specific design that was conceptually developed for a multi-chamber stack was proposed, which is based on the key concept of having enough rigid polymeric layers to be sealed by means of O-rings. Each MDC pilot plant is comprised of one stack of 15 MDC pilot units, each with $0.4\ m^2$ electrode area, housed in a 40-ft container with the rest of the peripheral elements. After acquiring knowledge from the first MDC pilot plant, the end plates were redesigned for the second MDC in order to improve the chemical compatibility and avoid deflection.

Chapter 5 will present the construction of this technology for its validation in two demo sites: Demo Site 1, the BWDP Racons and Demo Site 2, SWDP Fonsalia Guía de Isora, Spain, for brackish water and seawater desalination, respectively.

The full design of the MIDES pilot plant not only includes the engineering of the whole microbial desalination cell (MDC), but also a reverse osmosis (RO) pilot plant, wastewater pre-treatment, brackish water and seawater pre-treatment, and RO post-treatments (remineralization and disinfection). The following chapter will present the construction and operation of all the pilot plants designed in the MIDES project.

REFERENCE

Zhang F. and He Z. (2015). Scaling up microbial desalination cell system with a post-aerobic process for simultaneous wastewater treatment and seawater desalination. *Desalination*, **360**, 28–34.

Chapter 5

Construction and operation of the MIDES plant

Naiara Hernández-Ibáñez, Juan Arévalo, Vicente F. Mena, Victor Monsalvo-Garcia and Frank Rogalla

Aqualia, I+D Department, Av. del Camino de Santiago, 40, edificio 3, 4ª planta, 28050, Madrid, Spain

ABSTRACT

This chapter presents the construction, operation, and validation of all the MIDES systems, including water pre-treatment, wastewater pre-treatment, the microbial desalination cell (MDC), low-pressure reverse osmosis (RO), and post-treatment (remineralization and disinfection). MIDES technology has been validated with different water sources: brackish water from Demo Site 1, (Racons Brackish Water Desalination Plant (BWDP), located in Denia, Spain) and seawater from Demo Site 2 (Fonsalía Seawater Desalination Plant (SWDP), located in Guía de Isora, Spain). In this chapter, the preparation of both demo sites for the reception and installation of the pilot plants is also presented.

Keywords: Desalination, microbial desalination cell, pilot plants, water post-treatment, water pre-treatment

5.1 SYSTEM COUPLING

In water desalination processes, water pre- and post-treatment steps are necessary for stable operation of the desalination process and to meet the product water

doi: 10.2166/9781789062120_0105

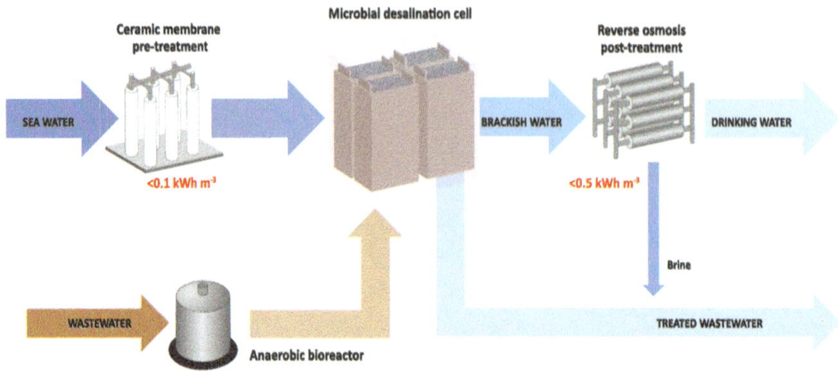

Figure 5.1 MIDES concept scheme. Source: Author's own work/property.

quality guidelines for drinking water, agricultural and industrial water. In this sense, MIDES technology includes the development of a series of pilot plants for saline and wastewater pre-treatment prior to the microbial desalination cell (MDC) pilot plant, as well as for the post-treatment of reverse osmosis (RO) permeate.

These pilot plants are not all operationally independent, but are coupled together in series and in parallel according to the above scheme (Figure 5.1). As a first step, brackish water and seawater are pre-treated in a microfiltration/ultrafiltration (MF/UF) membrane pilot plant. The permeate of the MF/UF pre-treatment step is stored in a tank which is then fed to the MDC pilot plant. At the same time, wastewater is treated in an anaerobic digester to ensure that the wastewater composition meets the operational guidelines, in terms of chemical oxygen demand and volatile fatty acids level (COD and VFAs), as this stream is used to power the MDC unit. The saline water stream that was pre-treated in the MF/UF step is then fed to the MDC pilot plant where it is partially desalinated. Subsequently, the MDC permeate is totally desalinated in the RO pilot plant. Finally, the RO permeate goes through the post-treatment stages of remineralization and disinfection.

For the design, construction, and coupling of each pilot plant, both the flow rates and the effluent water quality required in each subsequent treatment step have to be taken into consideration.

5.2 PERIPHERAL PLANT CONSTRUCTION

According to the final design, the MIDES pilot plant requires different peripheral plants to support the MDC operation. The MIDES plant comprises several different steps and water flows, from the pre-treatment of saline streams (with ceramic ultrafiltration) to the remineralization and disinfection pilot plants, which are required for post-treatment after the desalination step. A brief explanation is provided below of the different supplementary equipment and pilot plants required to run an MDC pilot plant.

5.2.1 Anaerobic digestion pilot plant for wastewater pre-treatment

To power the MDC unit, an anaerobic digestion unit (Figures 5.2 and 5.3) was used to pre-treat wastewater. Since VFAs, especially short chain ones, are a good fuel to power the MDC, it is necessary to maximize the concentration of VFAs with respect to the COD value. From the previous studies at pre-pilot scale a VFAs level higher than 300 mg/L was required for the MDC operation. Real wastewater was obtained from a nearby wastewater treatment plant (WWTP) in order to ensure that the wastewater composition met the operational requirements of the MDC system.

Anaerobic digestion is the first step in the wastewater pre-treatment scheme and is required to enrich the stream in VFAs – the fuel for MDC electroactive bacteria. The wastewater is first mixed and aerated inside a homogenization tank. The following step is to promote conditions (35°C) to generate acetogenesis under anaerobic conditions. As a safety measure, the produced biogas is measured (via a flow meter) and exits the pilot plant via an anti-flame device. The capacity of the plant is from 1–4.5 m^3/h of treated wastewater enriched in VFAs.

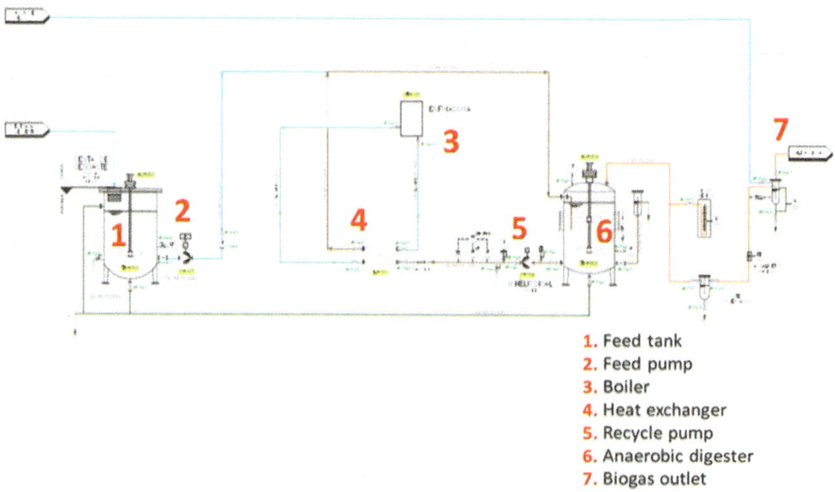

Figure 5.2 Piping and instrument diagram (P&ID) of the anaerobic digestion pilot plant. Source: Author's own work/property.

Figure 5.3 Anaerobic digestion pilot plant. Source: Author's own work/property.

The anaerobic digestion pilot plant was mounted on a skid and comprises the following elements:

- Metal frame to ensure the structural stability of the skid
- Homogenization/feed tank
- Anaerobic digester
- Two vertical mechanical stirrers
- Boiler
- Control and monitor panel
- Feeding and recirculation pump

5.2.2 Pilot plant for the pre-treatment of saline water

Saline water to be used as the feed stream to an MDC pilot plant must be pre-treated to ensure that the water quality is acceptable for the operating conditions of the MDC stack. These criteria are the same as for other membrane-based plants such as reverse osmosis or electrodialysis, with an silt density index (SDI) lower than 3 and turbidity lower than 1 NTU (nephelometric turbidity unit) being required.

Table 5.1 shows the characteristics of the membrane installed in the pilot.

Table 5.1 Ceramic membranes characteristics.

Membrane Characteristics	
Pilot plant production capacity	2.5–12.5 m^3/h
Membrane material	Ceramic (alumina)
Membrane form	Inner pressure monolithic
Nominal pore size	0.1 μm
Membrane area	25 m^2

The ceramic membranes are pressurized and operate in inside-outside filtration mode. The vertically mounted pressure vessel, containing the ceramic membranes, is presented below (Figure 5.4).

Figures 5.5, 5.6, and 5.7 show the P&ID, the layout and photograph of the ceramic pilot plant designed, constructed, and operated for the pre-treatment of the feed stream of the MDC pilot plant.

5.2.3 Reverse osmosis (RO) pilot plant

Reverse osmosis desalination, after the MDC, is the most important treatment to complete the desalination process which was partially achieved in the microbial

Figure 5.4 Ceramic membranes housed inside a pressure vessel. Source: Author's own work/property.

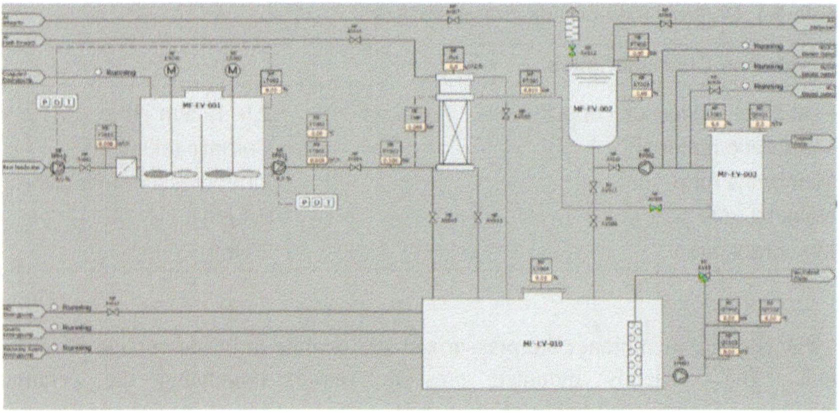

Figure 5.5 P&ID of the ceramic membrane pilot plant. Source: SMAS. Skywater Medioambiente y Servicios S.L.

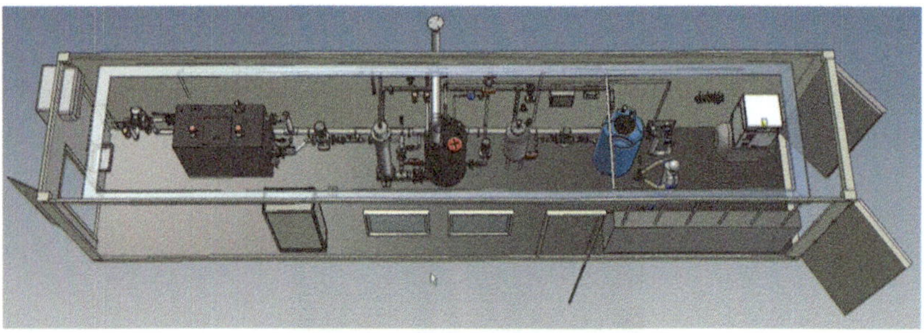

Figure 5.6 Layout of the ceramic membrane pilot plant. Source: SMAS. Skywater Medioambiente y Servicios S.L.

desalination cell. One of the main objectives of the RO pilot plant was to keep the energy consumption below 0.5 kWh/m^3.

On the other hand, the RO pilot plant had to be designed and built according to the Spanish standard RD140/2003 based on the water quality guidelines for human consumption in Spain. The elements of the pilot had to comply with all regulations:

- Raw water tank
- Automatic pH adjustment
- Chemical dosage (antifouling/reducing agent)
- Microfiltration cartridges (5 and 25 μm)
- High pressure pump
- Reverse osmosis system (vessels and membranes)
- Flushing pump for chemicals

Figure 5.7 Outside view of the ceramic membrane pilot plant container. Source: Author's own work/property.

A larger RO pilot plant was also designed (see Figures 5.8, 5.9 and 5.10), under the standard RD140/2003, with a lower energy consumption, and equipped with high quality RO membranes.

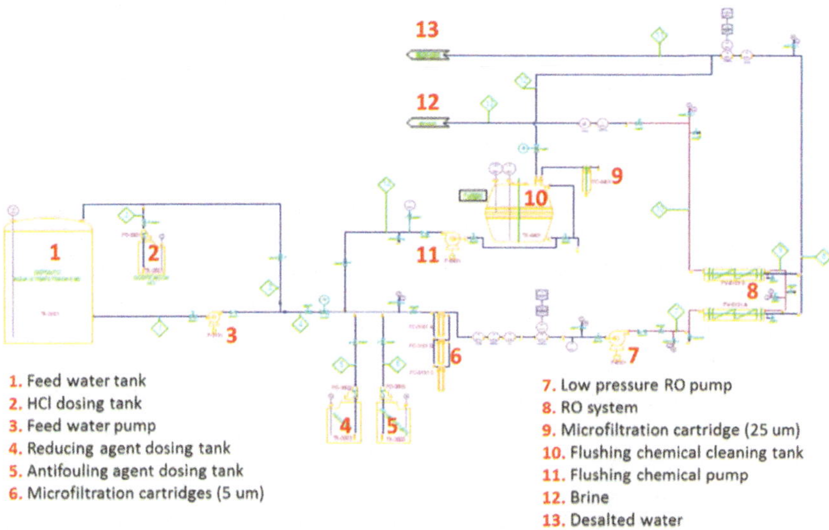

Figure 5.8 P&ID of the RO pilot plant. Source: RWB Water treatment.

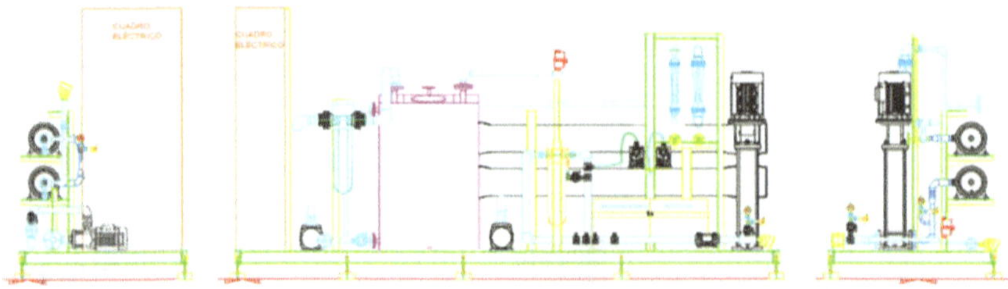

Figure 5.9 Layout of the RO pilot plant. Source: RWB Water treatment.

Figure 5.10 RO pilot plant. Source: Author's own work/property.

5.2.4 Remineralization

According to the Spanish standard RD140/2003 for water for human consumption, after any desalination process, the remineralization step is mandatory to ensure the water provided meets the water quality guidelines required by the sanitary authorities, not only for human health but also to minimize the interaction with the piping in the water transport and distribution system. After remineralization, the produced water is in effect adequate for human consumption according to Spanish standards.

The remineralization pilot plant designed for the MIDES project (Figure 5.11), consists of two main steps: acidification with CO_2 and calcite filtration.

The designed remineralization pilot plant is able to treat up to 0.72 m^3/h using calcite chips (average diameter: 1.5–3 mm) (see Figure 5.12). Before remineralization, the permeate is stored in a tank with the possibility to heat 1 m^3 of RO permeate up to 30°C in order to evaluate the remineralization process with respect to the temperature and the final quality of the produced water.

Figure 5.11 Remineralization pilot plant. Source: Author's own work/property.

Figure 5.12 Calcite used for percolation inside the columns. Source: Author's own work/property.

5.2.5 Disinfection with chlorine dioxide (ClO_2)

In many countries, remineralization is not the final step to produce water for human consumption. In several European countries a final disinfection step, prior to water transport and distribution, is mandatory. In Spain, chlorine disinfection is required (standard RD140/2003). To treat water produced by the MIDES pilot plant, ClO_2 was selected as the disinfection reagent. Its main advantages are the absence of chlorinated compounds as by-products; the preservation of the organoleptic water characteristics; and that no carcinogenic compounds are expected to be generated.

To generate (in situ) ClO_2, $NaClO_2$ and HCl (see reaction in Figure 5.13) are dosed by a pump in stoichiometric concentrations and mixed in a vacuum reaction chamber with a water stream to create a gas-water mixture. As a result of these conditions, ClO_2 is generated and added to the main remineralized water stream. Figure 5.14 shows the ClO_2 disinfection dosing systems.

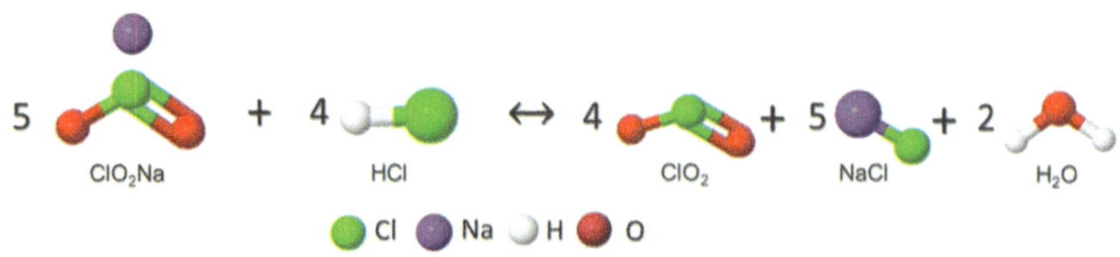

Figure 5.13 Reaction for the ClO_2 on site generation with HCl. Source: Author's own work/property.

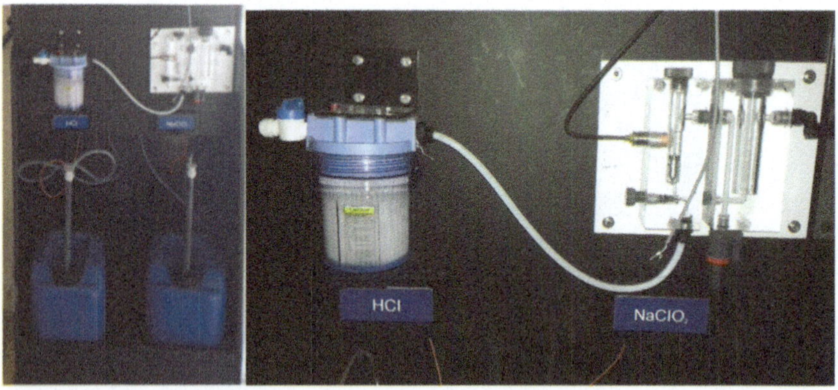

Figure 5.14 ClO_2 disinfection dosing systems. Source: Author's own work/property.

5.3 MDC CONSTRUCTION – LESSONS LEARNED

Following the ideas expressed in the design, the construction of the MDC stack and peripheral equipment/elements was started. The two MDC pilot plants were designed, constructed and installed in exactly the same way.

5.3.1 MDC stack construction

The MDC stack comprises 15 cells, and each cell has three chambers (anodic chamber/saline chamber/cathodic chamber). This distribution is repeated in all 15 cells constructed (see Figures 5.15 and 5.16).

The elements previously described are installed in each cell. Current collectors and carbon felts are installed in anodic and cathodic compartments, and each compartment (anodic, saline, and cathodic) is separated by ion exchange membranes (anionic for the anodic compartment (AEM), cationic in the cathodic compartment (CEM)).

The experience obtained assembling pre-pilot cells and especially the stacks was very valuable in the construction of the pilot-plant-size cells and stack. The design and construction of the pilot-scale cells is based on the knowledge generated in previous steps, and the valuable experience obtained gave the consortium

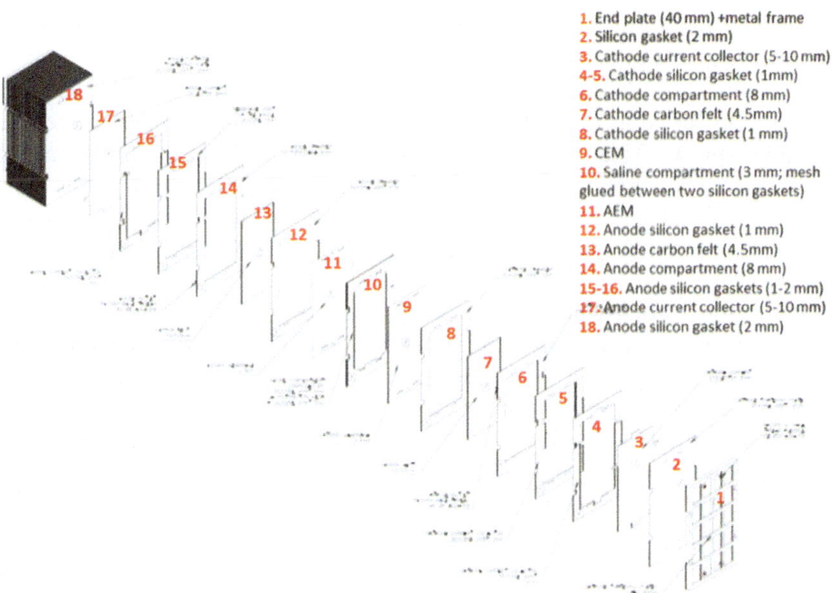

Figure 5.15 Isometric exploded view drawing of MDC pilot unit. Source: Hysytech S.r.l.

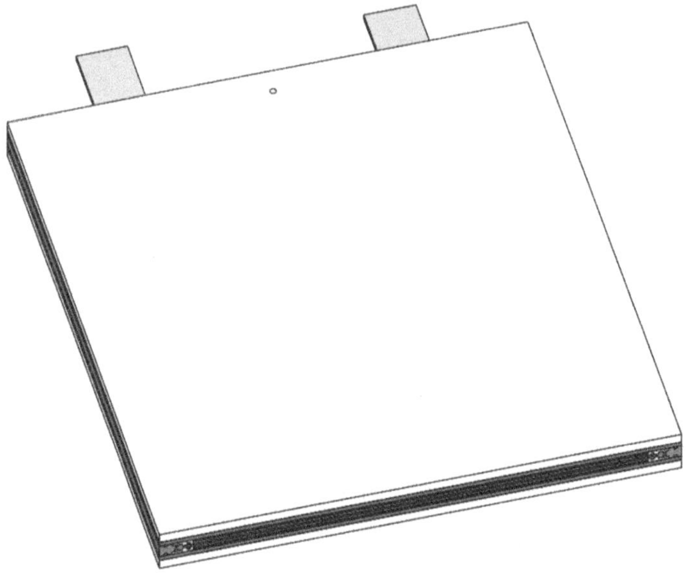

Figure 5.16 Section view of the MDC cell assembled. Source: Hysytech S.r.l.

members the confidence to overcome the challenges faced during the design and construction of the pilot.

To certify the design and gain more experience, two single cells were initially constructed and assembled in order to learn how to assemble the cells in the pilot plant. This process was documented and used to prepare a standard protocol for assembling the larger pilot plant. These two cells were mounted with all internal elements (see Figure 5.17) and hydraulic tests were performed to check for leaks, over pressure and possible failure spots in the cells (Figure 5.18).

The results obtained in this first proof of concept indicate good mechanical stability, with no damage to the membrane nor other internal elements of the cell. The hydraulic tests show promising results, with no leaks or mass transport problems between chambers (Table 5.2).

Following these promising results in the proof-of-concept phase, a full-scale MDC stack was constructed.

The full-scale pilot stack comprised 15 cells, which had to be assembled one by one, following the same protocol designed during the proof-of-concept phase involving construction of an individual cell. In Figure 5.19, the step-by-step procedure for the assembly of a single cell can be seen. The assembly of a single cell starts with the saline compartment (Figure 5.19(a)–(c)), followed by the anodic compartment assembly (Figure 5.19(d)–(i)). Then, the set of components

A) End plate
B) Insulating plate
C) Cathodic current collector
D) Cell frame
E) Cathode O-ring frame and carbon felt
F) Cathode flow distributor
G) Cationic exchange membrane
H) Saline flow distributor
I) Anionic exchange membrane
J) Anode flow distributor
K) Anode carbon felt
L) Anode current collector
M) Insulating plate
N) End plate
Ñ) MDC pilot unit assembled

Figure 5.17 Pilot plant cell units, step by step. Source: Author's own work/property.

Figure 5.18 Mechanical and hydraulic testing in the single cell pilot scale MDC. Source: Author's own work/property.

Table 5.2 Hydraulic test results.

Assembly	Test	Result
1 MDC pilot unit	Water flow 22.4 L/h Torque: 5 Nm	OK
2 MDC pilot units stack	Water flow 22.4 L/h Torque: 5 Nm	OK

Figure 5.19 Pictures related to the step-by-step assembly of a single cell. (a) Insulating End Plate, (b) Cell Frame, (c) Saline Compartment Flow Distributor (2 units), (d) Anionic Membrane, (e) Anodic O-ring Frame, (f) Anodic Flow Distributor, (g) Anodic Carbon Felt, (h) Anodic Current Collector, (i) Blind Insulating End Plate, (j) and (k) turn of the cell, (l) Insulating End Plate removal, (m) 2nd Saline Flow Distributor removal, (n) Cationic Membrane, (ñ) Cathodic O-ring Frame, (o) Cathodic Flow Distributor, (p) Cathodic Carbon Felt and, (q) Cathodic Current Collector. Source: Author's own work/property.

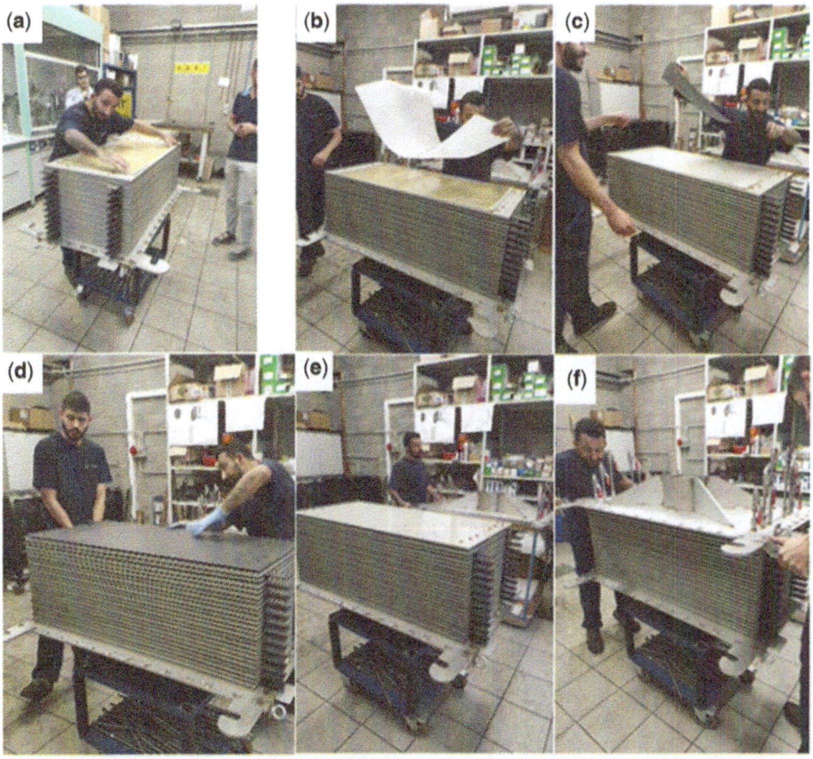

Figure 5.20 Pictures of the assembly of the cathodic compartment onto previously assembled cells (a) Cationic Membrane, (b) Cathodic Flow Distributor, (c) Cathodic Carbon Felt and, (d) Cathodic Current Collector, (e) Insulating End Plate, (f) End Plate. Source: Author's own work/property.

assembled previously is turned so that the saline compartment is located at the top (Figure 5.19(j)–(m)). Thereafter, the insulating end plate is removed to start the assembly of the cathodic compartment components as shown in Figure 5.19(n)–(q).

The cells were constructed one by one (see Figure 5.20). After each cell was assembled, hydraulic and mechanical testing was performed, to evaluate if the cell was correctly assembled and to ensure the integrity of the stack.

The hydraulic test was performed by feeding a 0.1 M NaCl solution to the stack at a flow rate of 150 L/h for 6 hours. Figure 5.21 shows a diagram of the proposed system and Figure 5.22. shows the MDC stack and tanks employed during the hydraulic tests. The liquid transfer was checked by the liquid level in the tanks.

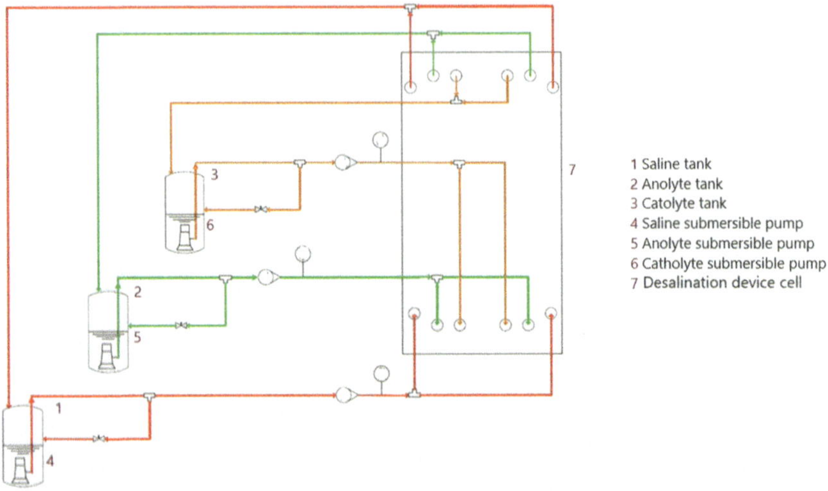

Figure 5.21 P&ID for the proposed MDC test system. Source: Hysytech S.r.l.

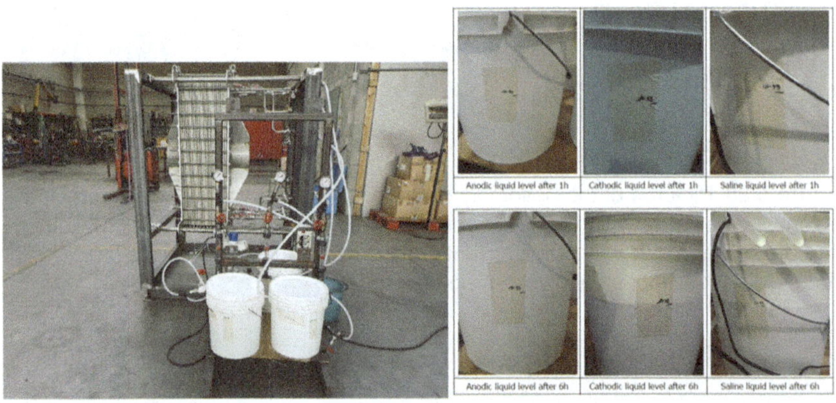

Figure 5.22 Pictures of the MDC stack and the tanks employed during the hydraulic tests. Source: Author's own work/property.

Throughout the test, where leak problems between chambers were observed, the last cell was disassembled, the components were checked, and the cell was reassembled and tested again.

As was described in chapter 4, the redesign of the endplates for the construction of the second MDC was necessary. Figure 5.23 presents pictures of the endplates (left) and connections of the cathodic line (right) for the second MDC stack.

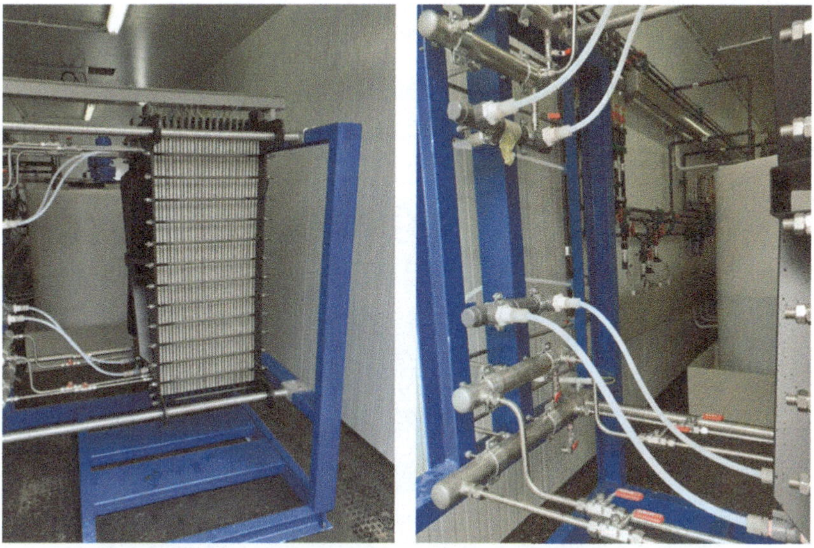

Figure 5.23 Pictures of the endplates (left) and connections of the cathodic line (right) for the second MDC stack. Source: Author's own work/property.

5.4 MDC REQUIREMENTS

5.4.1 MDC pilot plant peripheral elements

After the assembly of the MDC cells in the pilot, the stack was integrated with the rest of the peripheral pilot plant elements in a 40-ft. container as was described in Chapter 4.

The installation consists of two polymeric tanks with a volume of 1 m^3 each (anolyte and catholyte) and another polymeric tank with a volume of 0.35 m^3 (for sea/brackish water). The anolyte and saline streams are driven by 2 progressive cavity pumps, each stream can be pumped individually and/or simultaneously. The catholyte stream is corrosive ($FeCl_3$), and that is the reason why two plastic dosing pumps with a simple mechanical-controlled membrane pump were proposed (Figure 5.25(c)) instead of cavity pumps. To ensure correct MDC operation, both the anolyte and catholyte streams need have a homogenous composition, hence, two mechanical stirrers were in the anolyte and catholyte tanks.

After the integration of the peripheral elements, the factory acceptance tests (FATs) were performed where all the mechanical components of the first MDC pilot plant were examined (Figures 5.24 and 5.25). Firstly, water recirculation to the tank for each of the three streams was tested in order to check performance of pumps, mixers, and gas systems, as well as to identify possible leaks. The next

Figure 5.24 Picture of the MDC pilot plant container during integration of the peripheral elements of the installation. Source: Author's own work/property.

step was the water recirculation test through the stack. In this step, the stability of the flow rates of each stream was checked, at both the entry and exit of the stack, using the six rotameters installed (Figure 5.25(e)).

5.4.2 Control and monitoring system

The electric system installation was performed in parallel with the integration of the peripheral elements taking into consideration the knowledge obtained at the pre-pilot scale.

The automation, control and supervision systems are made up of a series of equipment, related to each other to allow correct operation of the pilot. The main element is the programmable logic controller (PLC) or automaton. This receives inputs and outputs, both digital and analogue, and uses these to control the process. Figure 5.26 shows the indoor electric panel. In order to view the process, a display screen is included that shows the information from the PLC and allows the user to interact with it (Figure 5.27).

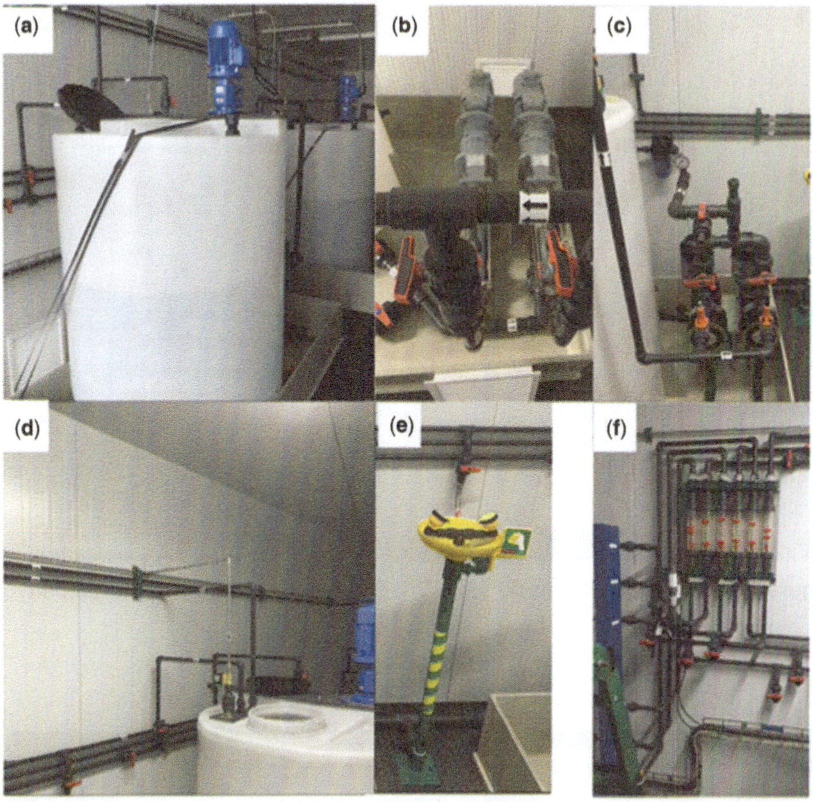

Figure 5.25 Pictures of some of the peripheral elements installed in the MDC pilot plant. (a) tanks, (b) saline and anolyte streams pumps, (c) catholyte stream pumps, (d) pipes and gas line, (e) eyebath and (f) rotameters. Source: Author's own work/property.

As in the case of the pre-pilot, the selected key variables for the control of the process were:

- pH of the anolyte, catholyte and saline tanks.
- Temperature of the anolyte, catholyte and saline tanks.
- Redox value of the anolyte, catholyte and saline tanks.
- Conductivity of the anolyte, catholyte and saline tanks.
- Anolyte potential.
- Catholyte potential.
- Cell potential of each cell.
- Current that flows in each cell.

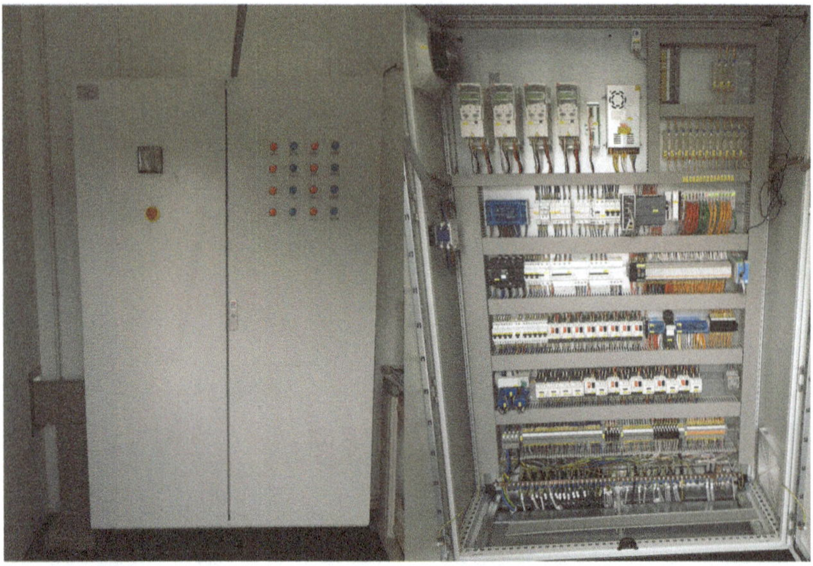

Figure 5.26 Electric panel inside the MDC. Source: Author's own work/property.

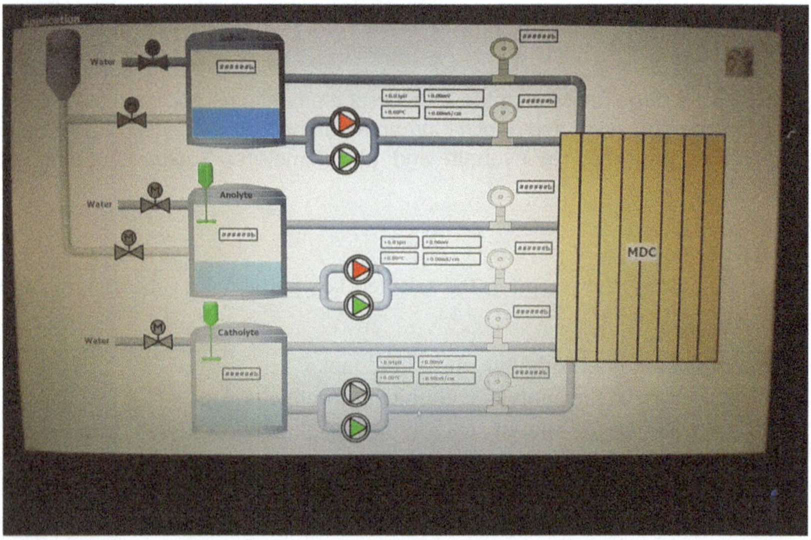

Figure 5.27 Display screen to operate the MDC. Source: Author's own work/property.

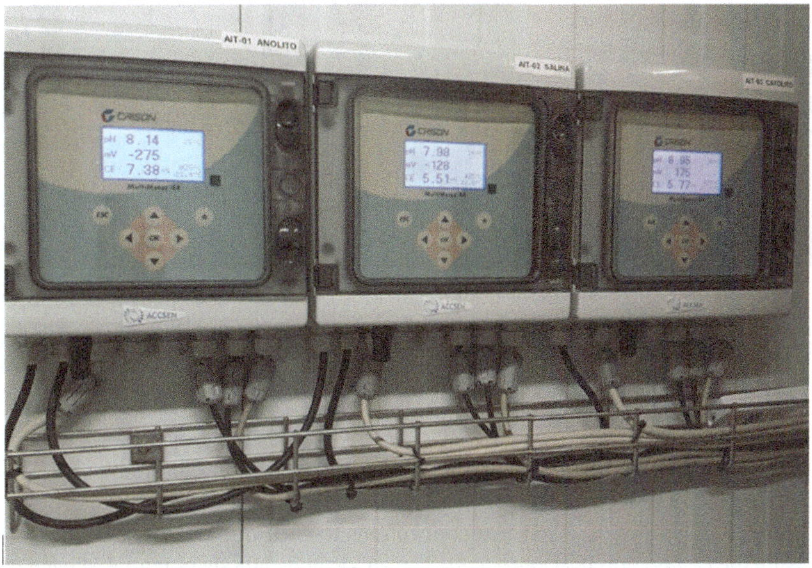

Figure 5.28 Picture of the multimeters installed in the MDC pilot plant. Source: Author's own work/property.

Figure 5.29 Electrical connections of the MDC pilot stack. Source: Author's own work/property.

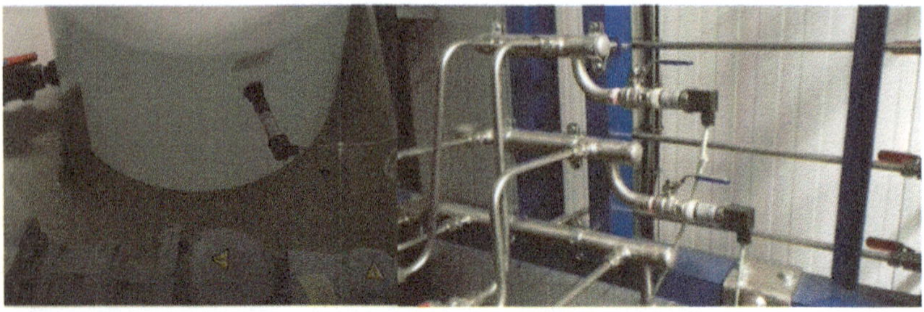

Figure 5.30 Image of the level transmitter located in the tank (left) and pressure transmitters installed in the inlet and outlet of the stack (right). Source: Author's own work/property.

Temperature, pH, redox, and conductivity values were monitored using three multimeters for the saline, anolyte and catholyte streams (Figure 5.28).

Figure 5.29 shows the electrical connections to each of the current collectors to measure the cell potential of every cell. The anode and cathode potential of each cell was also measured using AgCl/Ag reference electrode sensors that were located in the feeding pipes.

In addition to the parameters mentioned above, tank volume levels and inlet stack pressure were also monitored. For this, pressure transmitters and level transmitters were also installed (Figure 5.30).

5.4.3 Installation of MDC-RO systems at demo locations

MIDES technology has been validated at two demo sites: the first one in Denia and the second one in Guía de Isora on the Canary Islands (Figure 5.31). Both locations were selected to prove the MIDES technology in different environments and with different water sources (brackish water and seawater).

After finishing the peripheral and control system installation, the MDC pilot plants were transported to the two demo sites in Spain Figure 5.32).

The first MDC pilot plant was installed at the Racons Brackish Water Desalination Plant (BWDP) (Denia), Demo Site 1 (see Figure 5.33) for brackish water treatment, while the second MDC pilot plant is located at Fonsalía Seawater Desalination Plant (SWDP) (Tenerife) for seawater treatment (Figure 5.34).

The MDC pilot plants were placed near the pre-treatment pilot plant. Once placed, the pilot plant hydraulic and electrical connections were realized. For the hydraulic connections, the saline line was connected to the different pre-treatment pilot plant effluents, so that the impact of brackish water and seawater pre-treatment could be studied in the water desalination process.

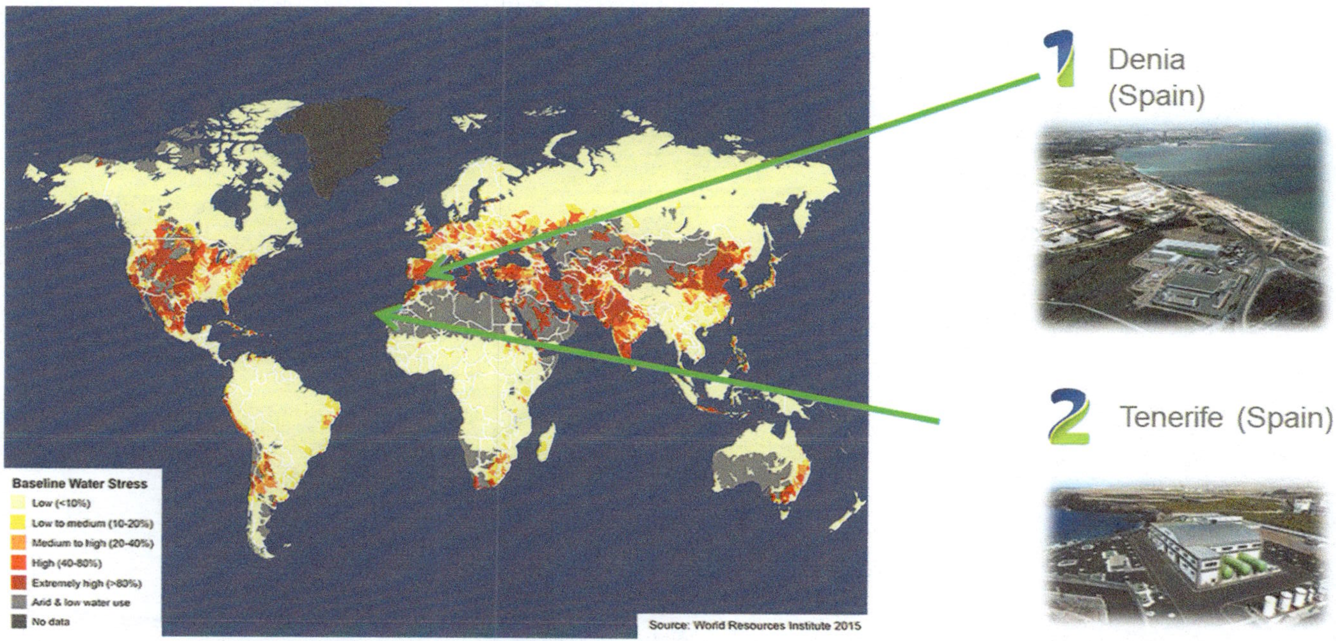

Figure 5.31 Demo location selected for the MIDES technology validation. Demo Site 1, Denia (Spain) and Demo Site 2, Tenerife (Spain). Source: World Resources Institute 2015, modified by the authors.

Figure 5.32 Placing the 40-ft. MDC pilot plant containers at (left) the Racons Brackish Water Desalination Plant (Denia) and (right) Demo Site 2 located at Fonsalía Seawater Desalination Plant (Tenerife). Source: Author's own work/property.

Figure 5.33 Images of the MDC pilot plant located at the Demo Site 1 (Denia). Source: Author's own work/property.

Figure 5.34 MDC final location at Demo Site 2 (Tenerife). Source: Author's own work/property.

5.5 PLANT OPERATION AND RESULTS

5.5.1 MDC operation and results

5.5.1.1 Abiotic tests

As in the case of the MDC pre-pilot, a series of abiotic experiments were performed with the full-scale pilot in order to corroborate its proper functioning as an electrochemical device. Firstly, a cell per cell electrode polarization test was performed in order to assure the correct electric connections of each electrode as well as a correct electrical circuit installation (Figure 5.35). After that, the polarization test of the electrodes of the whole cell was performed (Figure 5.36). Furthermore, a $NaHCO_3$ solution was recirculated through the stack to guarantee the conductivity of the liquid while an increasing potential from 0 to 1.2 V was applied to each cell using a power supply.

A preliminary process of conventional electrodialysis under abiotic conditions was performed in order to test if the system configuration was adequate for desalination operation. 600 L of $NaHCO_3$ solution (electrical conductivity (EC) = 6.7 mS/cm) was used as anolyte and catholyte and 175 L of 4 g/L NaCl solution as the saline stream ($EC = 7.5$ mS/cm). An electrical potential of 1.5 V was applied between the anode and cathode. The system was operated in batch mode.

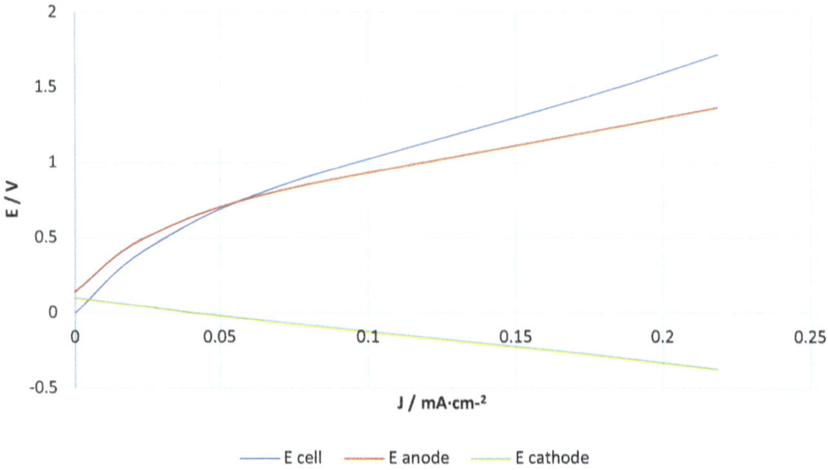

Figure 5.35 Polarization plot for cell 6 of the stack (area = 4000 cm^2). Source: Author's own work/property.

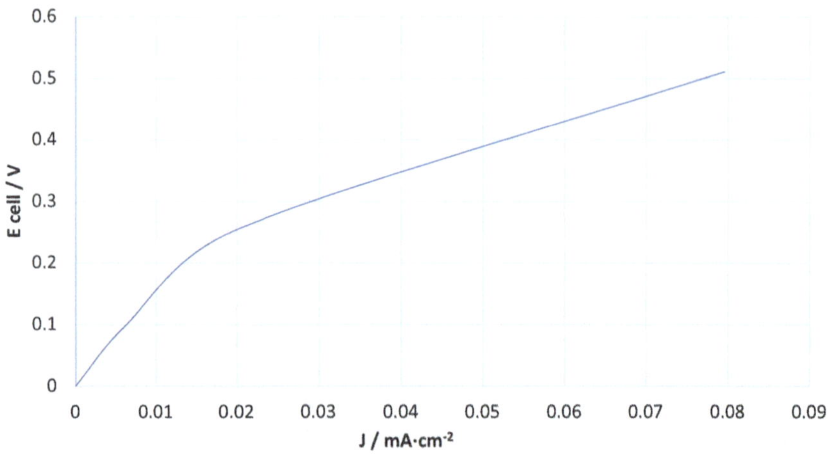

Figure 5.36 Polarization plot of all the cells in the stack. Area 4000 × 15 cm^2. Average current density. Source: Author's own work/property.

As can be observed in Figure 5.37 the salinity of the saline stream was decreasing during the process. Furthermore, the tank level remained constant for the three streams throughout the process. Both verifications indicate a correct component distribution inside the stack.

Figure 5.38 presents the current density produced during the desalination process. This value is similar to that obtained for the MDC pre-pilot stack. It

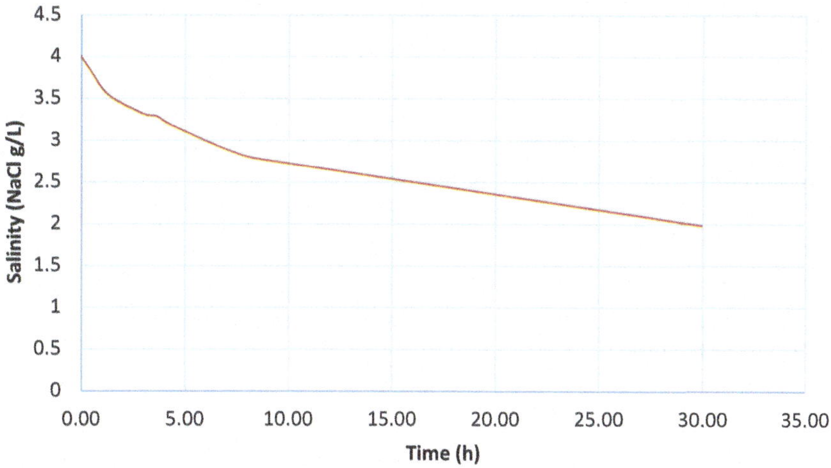

Figure 5.37 Salinity in the saline stream during the electrodialysis process under abiotic conditions. $E_{cell} = 1.5$ V. Source: Author's own work/property.

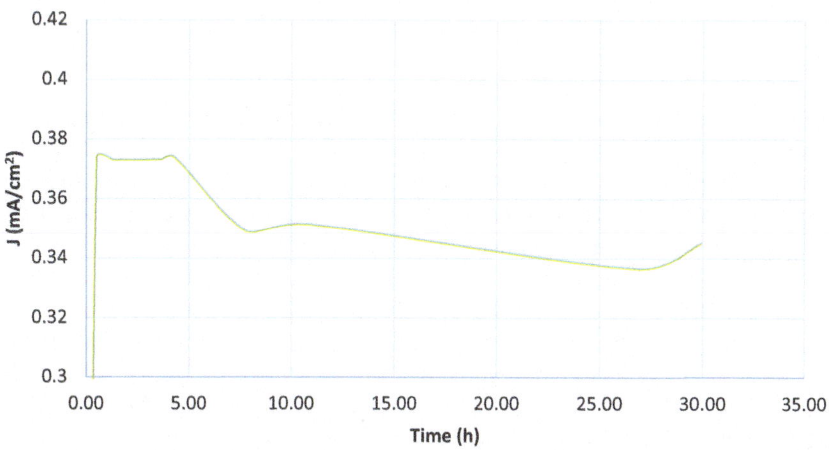

Figure 5.38 Current density produced during the desalination process under abiotic conditions. $E_{cell} = 1.5$ V. Source: Author's own work/property.

started at a maximum current density of 0.37 mA/cm^2 and was decreasing during the process, which is typical behaviour of conventional electrodialysis systems at a constant applied potential and operating in batch mode.

Table 5.3 presents the electrodialysis process parameters. The current efficiency (nc) obtained was lower than for the MDC pre-pilot stack, this is probably due to the

Table 5.3 Electrodialysis process parameters of the MDC pilot under abiotic conditions compared with parameters of the MDC pre-pilot stack.

Parameter	Units	Pre-pilot Stack	Pilot Stack
C_{si} $[NaCl]_i$	mol/m^3	73.5	68.9
C_{sf} $[NaCl]_f$	mol/m^3	8.5	34.3
nc	%	55.9	31.5
SEP	kWh/m^3	4.38	5.19
Nominal rate	L/m^2 h	0.8	1.18

C_{si} and Cs_f initial and final molar concentration of salt in the saline tank, respectively; SEP specific energy production.

high resistance of the cables from the power supply to the electrodes in the stack. This issue will not be a problem during the normal process operation under biotic conditions because a potential supplied by a power supply will not be required and the resistance of the cables will therefore not affect the current efficiency value. The rest of the parameters were in the order of what was obtained in the pre-pilot MDC, which indicates good scalability of the system as an electrochemical device.

5.5.1.2 MDC inoculation

Once the system was abiotically tested, the MDC inoculation process for the MDC start-up protocol was carried out. A total of 30 L of a pure culture of *Geobacter Sulfurreducens* strain supplied by IMDEA was used as the inoculum for inoculation into the anode compartment. Previously, the anolyte tank was filled with fresh water with the following mineral salts: 2.5 g/L $NaHCO_3$, 0.25 g/L NH_4Cl; 0.1 g/L KCl and supplemented with trace minerals and vitamin solutions. In addition, 10 L of inoculum of the operational MDC pre-pilot plant provided by IMDEA was added to the anolyte tank. Sodium acetate (3.28 g/L $NaC_2H_3O_2$) was used as the sole electron acceptor (see Figure 5.39). Before adding the *Geobacter* culture, a N_2 stream was gassed through the anolyte line in order to guarantee an anoxic environment in this line.

A volume of 600 L of $NaHCO_3$ solution ($CE = 5.5$ mS/cm) was used as catholyte and 175 L of 3.1 g/L NaCl solution as the saline stream ($CE = 5.7$ mS/cm). Then, the three streams were recirculated through the stack. Anode, cathode and cell potential, as well as the current intensity values, were recorded to control the biofilm development.

Figure 5.40 represents the anodic potential evolution during the inoculation process without an external potential during the anodic potential measurement. At the beginning of the inoculation process, the anodic potential was of 0.2 V vs AgCl/Ag, this potential is close to the graphite electrode open potential. As can

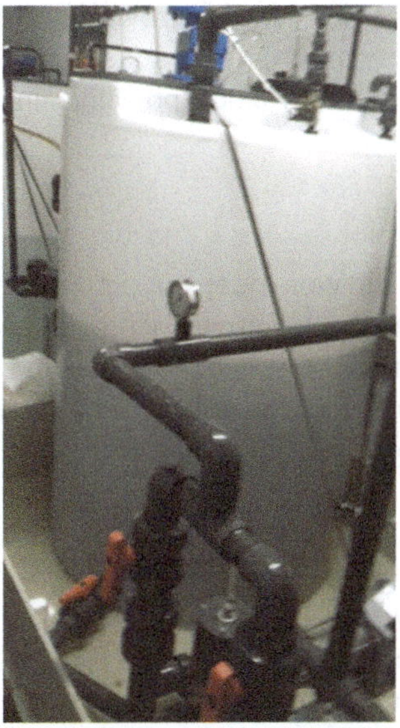

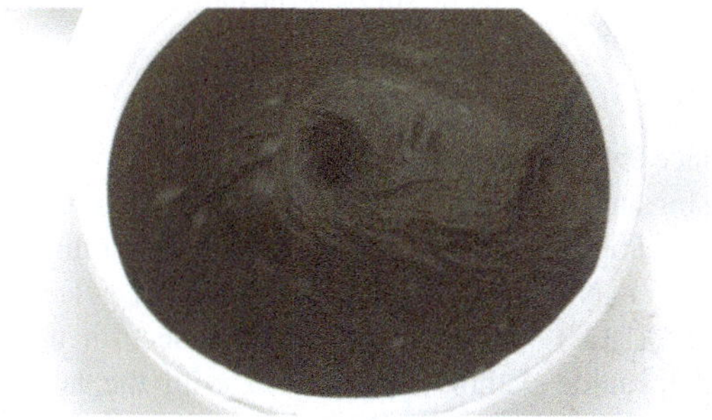

Figure 5.39 Anolyte solution during the inoculation process for the MDC start-up protocol. Source: Author's own work/property.

be observed, the anodic potential decreased throughout the inoculation process to an anodic potential of −0.7 V vs AgCl/Ag. This phenomenon is characteristic of *Geobacter Sulfurreducens* electroactive behaviour due to the charge accumulation

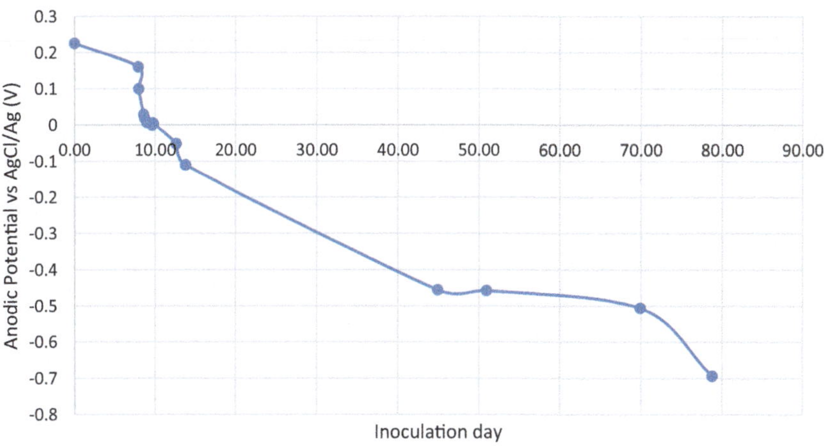

Figure 5.40 Potential for anode and cathode during the inoculation process for the MDC start-up protocol (measured at open circuit potential). Source: Author's own work/property.

after acetate oxidation (Borjas, 2017). This issue indicates biofilm growth in the anode of the stack. The same phenomenon was observed in the 15 anodes of the stack. The cell potential (i.e. 1.0 V) was maintained until anode potential and current density were almost constant.

5.5.1.3 Biotic desalination

Once the bioanode was considered stable, the first desalination cycle was performed using an NaCl solution as the saline stream with an initial conductivity of 4 mS/cm. The anolyte solution used consisted of a solution with the following composition: 2.5 g/L $NaHCO_3$, 0.25 g/L NH_4Cl; 0.1 g/L KCl and sodium acetate (3.28 g/L $NaC_2H_3O_2$) was used as the sole electron acceptor. To reduce operation cost, ferricyanide catholyte employed with the lab and pre-pilot scale MDC for the cathodic reaction ($Fe(CN)_6^{3-} + 1e^- \rightarrow Fe(CN)_6^{4-}$, $E^0 = 0.36$ V) was replaced by a $FeCl_3$ solution (60 mM $FeCl_3$) ($FeCl_3 + 1e^- \rightarrow FeCl_2$, $E^0 = 0.771$ V). The system was operated in batch mode with a flow rate of 300 L/h for anolyte and catholyte streams and 175 L/h for the saline stream. After each batch desalination experiment, the cathodic line was cleaned with an acidic solution in order to remove the possible precipitation of $FeCl_3$ in the cathodic chamber.

Figure 5.41 shows an example of the brackish water electrical conductivity evolution during a desalination cycle using the pilot MDC system from 4.1 to 0.946 mS/cm. As can be observed, it was possible to carry out the desalination of brackish water to drinking water without any energy supply (excluding pumping energy). Table 5.4 includes an example for brackish water desalination using the

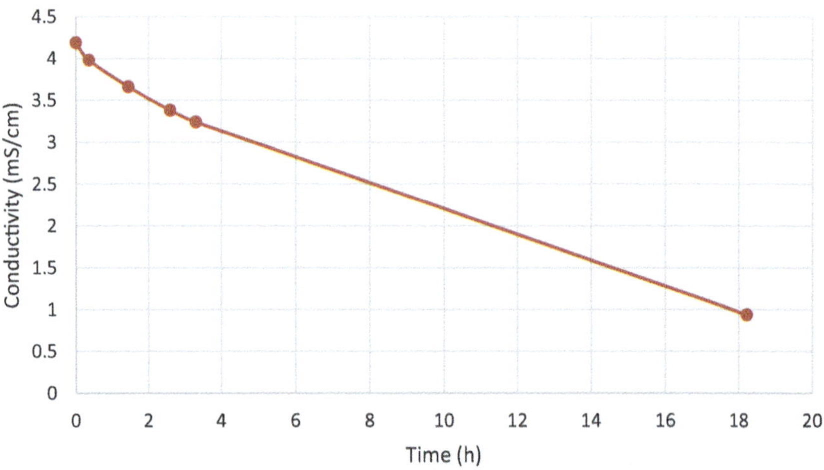

Figure 5.41 Salinity in the saline stream during a desalination cycle. Brackish water initial conductivity of 4 mS/cm); catholyte (0.06 M); synthetic anolyte. Source: Author's own work/property.

Table 5.4 Preliminary results for MDC pilot system for the desalination of synthetic brackish water.

Parameter	Units	Brackish Water
Initial conductivity saline compartment	mS/cm	4.2
Final conductivity saline compartment	mS/cm	0.946
Nominal desalination Rate	L/m^2 h	4.13
Salt removal	%	79.44
Specific electric production	kWh/m^3	0.867
Current Efficiency	%	89.0
Coulombic efficiency	%	20.47
Treatment Capacity*	g COD/m^3day	8386

*Considering the volume of anolyte compartment.

pilot MDC system at Demo Site 1. The experimental results show desalination nominal rate (NDR) of 4.13 L/m^2 h and COD removal rate of 8386 g COD/m^3day, these values are comparable to the laboratory and pre-pilot scale systems, showing good similarity for the future of the scale up of the MDC system.

REFERENCE

Borjas Z., Esteve-Núñez A. and Ortiz J. M. (2017). Strategies for merging microbial fuel cell technologies in water desalination processes: start-up protocol and desalination efficiency assessment. *Journal of Power Sources*, **356**, 519–528.

Chapter 6

Simulation and control systems in MIDES

Erhard Perz[1], Fatima Dargam[1], Stefan Bergmann[1], Ekaterina Rodionova[1], Pedro Sousa[2], Francisco Alexandre A. Souza[2] and Tiago Matias[2]

[1]*SimTech Simulation Technology, Graz, Austria*
[2]*Oncontrol Technologies, Coimbra, Portugal*

ABSTRACT

This chapter presents the work concerning the modelling and simulation of the overall MDC process, as well as its performance analysis and optimization. It also focuses on the support that the work brings for operational decisions on desalination plants, specifically applied to a microbial-powered approach for water treatment and desalination, starting from the stages of process modelling, process simulation, optimization and lab-validation, through the stages of plant monitoring and automated control. The work is based on the application of the environment IPSEpro from SimTech for the stage of process modelling and simulation; and on the system Databridge from Oncontrol for automated control, which employs techniques of machine learning.

Keywords: Automated control, Databridge, desalination plants, IPSEpro, machine learning, MDC, microbial desalination cell, operational decision support, plant monitoring, process modelling, process simulation.

doi: 10.2166/9781789062120_0137

6.1 INTRODUCTION

This chapter describes the part of the work carried out within the MIDES project relative to the modelling and simulation of the overall microbial desalination cell (MDC) process, as well as its performance analysis and optimization. Some focus concerning the support for operational decisions on desalination plants, specifically applied to a microbial-powered approach for water treatment and desalination is also exploited. The supported operational decisions and actions on the implementation of the desalination technology, follow a roadmap that starts from the stages of process modelling, process simulation, optimization and lab-validation, through the stages of plant monitoring and automated control. For the stage of process modelling and simulation, the developed work is based on the use of the IPSEpro environment (Perz *et al.* 1995; SimTech Simulation Technology; SimTech Simulation Technology; SimTech Simulation Technology), while for the stage of automated control, the Databridge system is applied, employing techniques of machine learning (MIDES Consortium).

Chapter 6 is structured in the following way: Section 6.2 presents the overall process modelling and simulation performed for the MIDES technology using IPSEpro. In Section 6.3, the energy consumption and optimization analysis for the MIDES solution is described. Section 6.4 introduces the implementation of the IPSEpro web-based platform that allowed the sharing of the simulated MDC overall process online among the MIDES partners for further analysis and potential optimization of the process. Section 6.5 presents the monitoring and automated control stages of the project (including the machine learning approach), at lab, pre-pilot and pilot scale. Section 6.6 discusses the operational decision making and decision support, given by the developments described in the previous sections, in the implementation of the MIDES pilot desalination plant. Finally, Section 6.7 draws some conclusions about the work carried out in MIDES as a whole and the impact of the work presented in this chapter.

6.2 OVERALL MIDES PROCESS MODELLING AND SIMULATION

In the MIDES project (Horizon 2020 Project MIDES; Zamora *et al.* 2019), the microbial cell design, engineering and testing was implemented in SimTech's IPSEpro simulation environment (SimTech Simulation Technology; SimTech Simulation Technology) and integrated in the MIDES customized library of component models (MIDES_Lib). Based on the components in the MIDES_Lib, a model of the overall MIDES process was developed. A simulation model of the overall process provided valuable insights at all stages of the project, supporting a basis for decisions concerning the integration of the MIDES process, and also for assessing the impact of changed component parameters due to up-scaling; as well as for analysing the performance of the pilot plant. The following section briefly presents the IPSEpro system, and its use and further development within the project.

6.2.1 IPSEpro – SimTech's integrated process simulation environment

IPSEpro (SimTech's Integrated Process Simulation Environment) is a heat balance and process simulation software package, which is currently one of the most comprehensive and versatile process modelling systems available. IPSEpro can be applied in a wide range of applications, within the areas of desalination; geothermal energy; refrigeration; concentrating solar power; and thermal power. From the early 1990s to present times, IPSEpro users have continuously been publishing their results obtained from the modelling, simulation, optimization and validation work using IPSEpro (among them Aneke *et al.*, 2011; Karellas *et al.*, 2012). With its various modules, IPSEpro supports users throughout the entire lifecycle of a process plant, from conceptual design to on-line plant performance monitoring and optimization, as shown in Figure 6.1. Due to its unique level of flexibility and its open architecture, IPSEpro is ideal for implementing customized modelling solutions (SimTech Simulation Technology; SimTech Simulation Technology; SimTech Simulation Technology).

In IPSEpro, process models are created using the Process Simulation Environment (PSE). In IPSEpro, the user sets up the process model graphically by drawing a flowsheet using components from a model library. Required data is

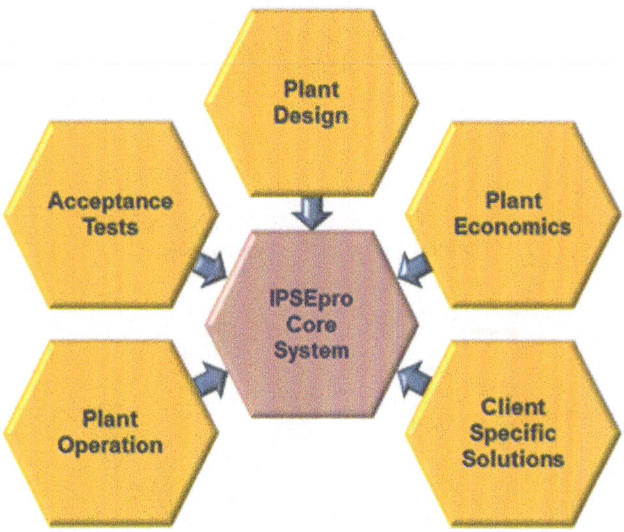

Figure 6.1 IPSEpro system capabilities (Perz *et al.*, 1995).

entered directly in the flowsheet. By drawing the flowsheet and entering the data, the user implicitly creates a system of algebraic equations, which is then solved by IPSEpro's solver core. Results are displayed graphically in the flowsheet. The modelling approach has been described in detail in (Perz *et al.*, 1995).

As IPSEpro is an open and flexible framework, component equations and physical property methods are not part of the core software. Instead, application-specific information is contained in model libraries, which can be created and modified using a special Model Development Kit (MDK). For instance, a model library specifically developed for ORC (Organic Rankine Cycle) processes (Perz & Erbes, 2011) includes a comprehensive set of component models based on working fluids used, as well as models for the part-load behaviour of the components, so that the user can analyse the off-design characteristics of ORC plants. The same applies for desalination plants, using a dedicated Desalination Model Library (SimTech Simulation Technology). Figure 6.2 illustrates the IPSEpro Simulation Package general architecture (SimTech Simulation Technology; SimTech Simulation Technology).

6.2.2 Main component models for the MDC process implemented in IPSEpro

IPSEpro Process Simulation Environment (PSE) (SimTech Simulation Technology)] and IPSEpro Model Development Kit (MDK) (SimTech Simulation Technology) were both used in the development of MIDES component models, with the creation of a customized model library called MIDES_Lib, as well as in the creation of the overall process model including the MDC. The component models were used to model the MDC and the MDC-RO (reverse osmosis) technology solution of the project. The basic concepts for the implementation of the MDC-RO solution in MIDES were taken from the principles explained in the sub-section which follows.

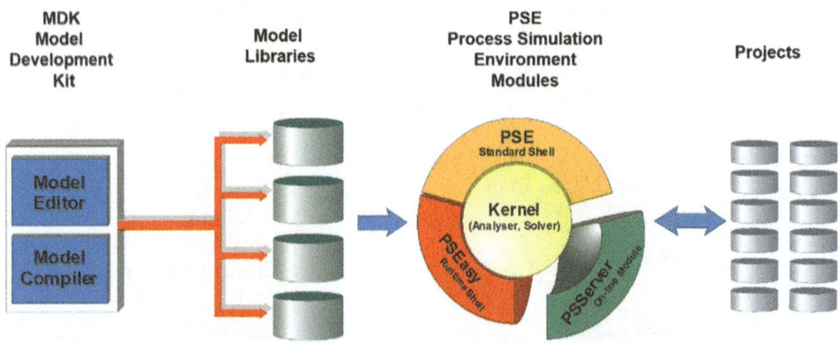

Figure 6.2 IPSEpro general architecture (SimTech Simulation Technology).

6.2.2.1 MDC-RO – MIDES solution

The microbial desalination cell (MDC) combined as a pre-treatment of the reverse osmosis (RO) desalination process comprises the main part of the underlying technology pursued in the MIDES project. MDC-RO and the overall MIDES solution are illustrated in Figure 6.3 and have been thoroughly described in Chapter 2.

According to the objectives of the MIDES project, the MDCs concurrently treat wastewater and generate energy to achieve desalination and are used in this context to overcome thermodynamic limitations of RO.

Most of the material concerning the MDC details found in this section comes from the MIDES project, from the work done on the 'Definition of the MDC mathematical model for the lab scale set-up operation' (MIDES Consortium). This step also supported the operational model of the full scale demo unit. In the early phase of model development, a dynamic model, which in parts originates from an electrodialysis model, was developed and used for time-dependent performance predictions. The model even predicts the dynamic response of the system output by changing the influent concentration of acetate, the external resistance, and the feed salt concentration.

From the early dynamic model, a simplified steady-state model based on the principal relationships of mass conservation was derived (MIDES Consortium). It additionally uses correlation equations fitted to lab and pre-pilot plant data and enables the calculation of performance figures for continuous operation of different process configurations, and can be used to study the impact of different parameters on the system performance to optimize the salt removal efficiency. The model is directly related to the data accuracy of the parameters which are given to represent the real operational conditions of the process. Limitations on the data accuracy can represent a model limitation in its validation.

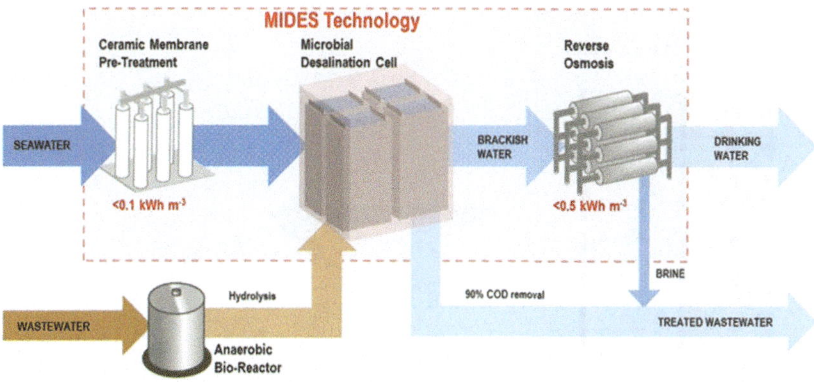

Figure 6.3 MIDES Technology including MDC-RO (MIDES Consortium).

The design and optimization of an MDC process must take into account several aspects that is., stack configuration, design of turbulence promoters, mode and conditions of operation, and characteristics of ion exchange membranes like perm selectivity, feed acetate concentration, flow, current or voltage, etc. In addition, other parameters related to the MDC bioanode (illustrated in Figure 6.4) performance must be taken into account: kinetics and thermodynamics, thickness of biofilm, biofilm electric resistance, and mass transport of the organic substrate in the bioanode. Figure 6.5 shows the integration of different simulation models for MIDES technology, considering the ceramic membrane pre-treatment, the MDC unit, and the RO post-treatment unit (MIDES Consortium).

For modelling MDC-RO in the IPSEpro simulation environment, some of the parameters shown in Table 6.1 were taken into consideration.

The implementation of the model library, MIDES_Lib, for modelling the MIDES solution, is based on the desalination model library that is available for IPSEpro. Desal_Lib contains: Process Components and Physical Properties methods to model sweet water production from saline water via thermal and membrane desalination processes. Only models that are relevant for membrane-based technology reverse osmosis (RO) were taken over from Desal_Lib.

In order to enable modelling of the MIDES solution, it was necessary to modify and extend:

- Stream models that represent the flows between the components of the system, in IPSEpro implemented as Connection models.
- Process components, models that represent the actual components in the system, in IPSEpro implemented as Unit models. It was also necessary to implement completely new unit models.

Figures 6.6–6.10 and 6.12 show the data input dialogs for various connection and component models available in MIDES_Lib. They are shown here to illustrate

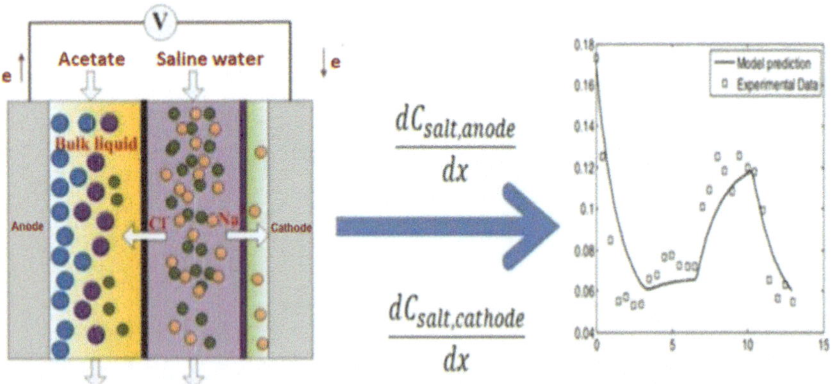

Figure 6.4 Concept of the MDC system (MIDES Consortium).

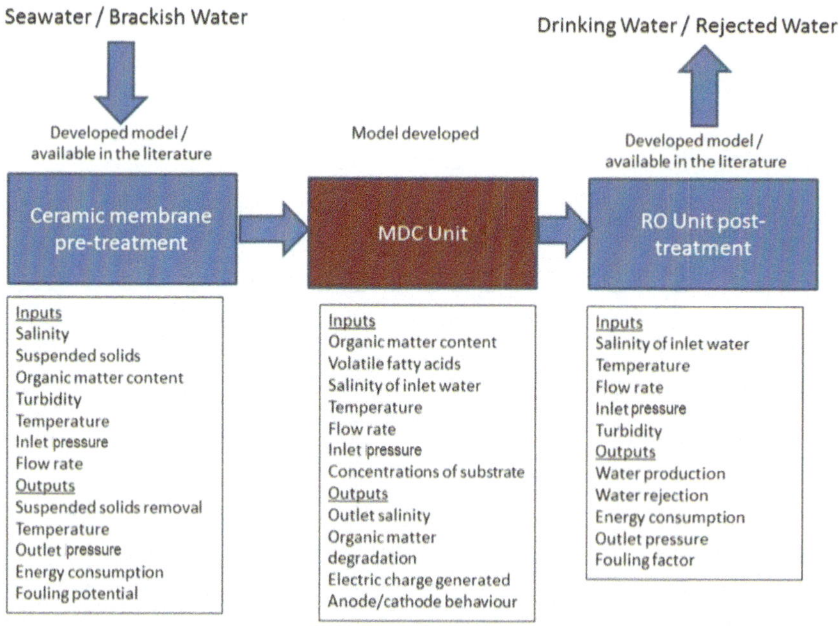

Figure 6.5 Flow chart of the integration of different simulation models for MIDES technology (MIDES Consortium).

which data items are available and accessible in the models. It is important to notice that the numeric values in the dialogs do not represent results of calculations, but inputs. Only in the case that they are marked as *set* do they represent fixed, prescribed values that will also show up in calculation results. A more detailed description of the meaning of the options in the data input dialog can be found in (SimTech Simulation Technology).

6.2.2.2 Extended connections for models

To include the MIDES specific details, which are transferred between different components, the connections had to be extended. The connection OW_Stream for representing streams of "organic wastewater" was developed new for the simulation in the IPSEpro environment.

The values that are presented in the subsequent figures originate from the modelling of the lab/pre-pilot MDC units and show typical estimate values in the data input dialogs. These values are not related to actual operating points from the MIDES plants.

6.2.2.2.1 Using wastewater in an OW_Stream

The organic wastewater stream connections OW_Streams contain wastewater with a variable content of organic matter which is transferred throughout the process.

Table 6.1 MIDES (MDC-RO) process – parameters analysis (MIDES Consortium).

#	Parameters	Description of the Parameter	MIDES Process
1	spec_E_el_DW	Specific electrical energy consumption for drinking water production [kWh el/m^3 DW]	Low <0.6 kWh/m^3 target
2	spec_E_th_DW	Specific thermal energy consumption for drinking water production [kWh th/m^3 DW]	No
3	spec_E_el_TWW	Specific electrical energy consumption for treated waste water [kWh el/m^3 TWW]	Low
4	spec_E_th_TWW	Specific thermal energy consumption for treated waste water [kWh th/m^3 TWW]	Medium
5	Exp_Life	Main components expected lifetime	High
6	rem_COD	Chemical Oxygen Demand removal [COD out/COD in]	High 90% target
7	rem_sal	Salinity removal [salinity out/salinity in]	Moderate
8	rec_DW	Drinking water recovery ratio [DW produced/SW feed]	50% target
9	E_el	Electrical energy use [consumption (+)/generation (−)]	Low
10	E_th	Thermal energy use [consumption (+)/generation (−)]	No
11	Chem_in	Chemicals consumption (various)	Medium
12	Air_in	Air consumption (for aeration)	No
13	Gas_out	Gas output (various)	Yes
14	Sludge_out	Sludge output	Low
15	RM_rec_out	Recovered Raw Materials output (various)	No
16	Chem_out	Spent chemicals output (various)	Moderate
17	Contam_in	Incoming contaminants requiring extra attention	Yes
18	Amb_p	Ambient pressure	Insignificant
19	Amb_t	Ambient temperature	Insignificant

20	Amb_hum	Ambient relative humidity	Insignificant
21	SW_flow_in	Seawater inlet flowrate	Low 3.5 L/min design value
22	SW_p_in	Seawater inlet pressure	Low
23	SW_t_in	Seawater inlet temperature	Significant
24	SW_sal_in	Seawater salinity	Significant 35 g/L typical
25	SW_SS_in	Seawater suspended solids	Significant 100–200 ppm
26	SW_OM	Seawater organic matter content	Significant
27	SW_turb	Seawater turbidity	Significant 100–300 NTU
28	SW_pH	Seawater pH	No significant influence
29	WW_flow_in	Wastewater inlet flowrate	Low 3 m^3/h target 7.0–28.0 L/min
30	WW_p_in	Wastewater inlet pressure	Low
31	WW_t_in	Wastewater inlet temperature	Medium
32	WW_COD_in	Wastewater inlet Chemical Oxygen Demand	Medium 1000–4000 mg/L
33	WW_BOD_in	Wastewater inlet Biochemical Oxygen Demand	Medium
34	WW_TSS_in	Wastewater inlet Total Suspended Solids	Low
35	WW_VSS_in	Wastewater inlet Volatile Suspended Solids	Low
36	WW_TN_in	Wastewater inlet Total Nitrogen	Medium
37	WW_NH4N_in	Wastewater inlet NH4-N	Medium
38	WW_TP_in	Wastewater inlet Total Phosphorus	Low
39	WW_Nut_rat	Wastewater Nutrient ratio BOD5:N:P	Low
40	Bri_flow_out	Brine outlet flowrate	Low
41	Bri_p_out	Brine outlet pressure	Low

(Continued)

Table 6.1 MIDES (MDC-RO) process – parameters analysis (MIDES Consortium) (*Continued*).

#	Parameters	Description of the Parameter	MIDES Process
42	Bri_t_out	Brine outlet temperature	Significant
43	Bri_sal	Brine salinity	Medium
44	Bri_SS_out	Brine suspended solids	
45	Bri_OM	Brine organic matter content	
46	Bri_turb	Brine turbidity	
47	Bri_pH	Brine pH	No significant
48	DW_flow_out	Drinking Water outlet flowrate	Low 150 L/h = 3.6 m^3/d target 1.2–5.0 m^3/d
49	DW_p_out	Drinking Water outlet pressure	Low
50	DW_t_out	Drinking Water outlet temperature	No significant
51	DW_sal	Drinking Water salinity	Low <0.5 g/L target
52	DW_pH	Drinking Water pH	Medium
53	DW_Standards	Drinking Water standards to be met	Significant
54	TWW_flow_out	Treated Wastewater outlet flowrate	Low
55	TWW_p_out	Treated Wastewater outlet pressure	Low
56	TWW_t_out	Treated Wastewater outlet temperature	Medium
57	TWW_COD_out	Treated Wastewater outlet Chemical Oxygen Demand	Low <100 ppm COD target
58	TWW_BOD_out	Treated Wastewater outlet Biochemical Oxygen Demand	Low
59	TWW_TSS_out	Treated Wastewater outlet Total Suspended Solids	Low
60	TWW_VSS_out	Treated Wastewater outlet Volatile Suspended Solids	Low
61	TWW_TN_out	Treated Wastewater outlet Total Nitrogen	Low

62	TWW_NH4N_out	Treated Wastewater outlet NH4-N	Low
63	TWW_TP_out	Treated Wastewater outlet Total Phosphorus	Low
64	TWW_Standards	Treated Wastewater standards to be met	Significant
65	Foul_be	Fouling behaviour	Significant
66	Scal_be	Scaling behaviour	Significant
67	Pre_req	Pretreatment requirements	
68	Cond_req	Conditioning requirements	
69	O&M	O&M related parameters (cleaning, maintenance, replacement etc.)	Significant

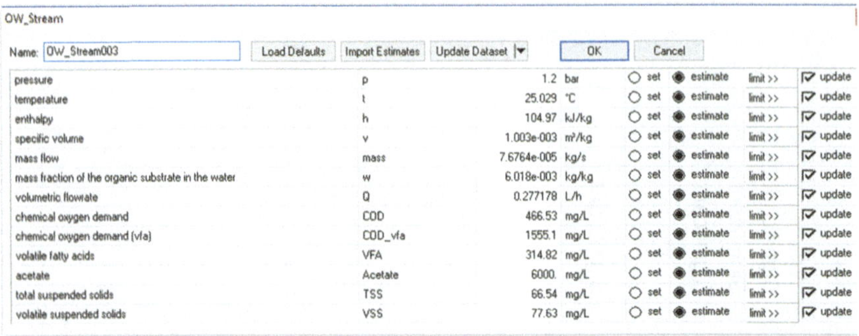

Figure 6.6 OW_Stream data input dialog for wastewater quality of incoming stream (MIDES Consortium).

Wastewater being used as fuel for MDCs has been discussed in several recent presentations from work package partners, as given in (Arevalo, 2018; Ortiz *et al.*, 2018; Zamora *et al.*, 2018) and also in previous chapters (e.g., Chapter 2).

The variables describing the physical properties of wastewater and selected key indicators for the organic load in an OW_Stream are shown in Figure 6.6.

If the wastewater enters the process via an OW_Source (organic-wastewater-source model connection), it allows either the selection of different wastewater specification data or the introduction of a user-defined specification directly in the respective OW_Stream connection. Typical specifications for domestic wastewater, domestic wastewater with molasses or brewery wastewater are available. These specifications can be selected via a dropdown list in the OW_Source as shown in Figure 6.7.

6.2.2.2.2 Saline water streams

In terms of saline water streams connection-extensions for the model, the basic connection SW_Stream and two more detailed representations, SW_Stream_ext1

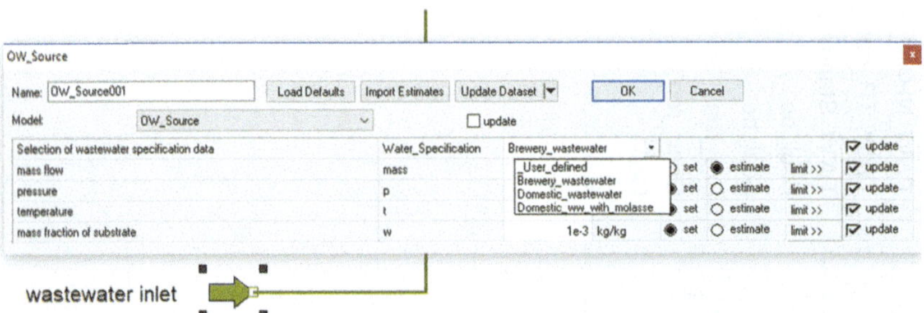

Figure 6.7 Using wastewater selection in OW_Source (MIDES Consortium).

and SW_Stream_ext2, were considered. SW_Stream_ext1 includes a very detailed description with individual ionic species, while SW_Stream_ext2 includes selected key indicators for saline water. The data input dialogs for these connections are shown in the following sub-sections.

6.2.2.2.3 SW_Stream

The purpose of the saline water stream connection SW_Stream is the transfer of saline water with a variable salt mass fraction. The SW_Stream (Figure 6.8) is the standard connection for all components where saline water is used.

6.2.2.2.4 SW_Stream_ext1

The purpose of SW_Stream_ext1 (saline water stream connection-extension1) is the transfer of saline water where the ionic species present in the water are of interest. Hence the standard seawater stream has been extended with the prevailing ionic species. The extension and the list of ionic species considered is shown in Figure 6.9.

6.2.2.2.5 SW_Stream_ext2

A second saline water stream connection-extension SW_Stream_ext2 was considered for the transfer of mass and several selected key indicators for saline water. This is shown in Figure 6.10.

6.2.2.3 RO – reverse osmosis models

The saline water reverse osmosis SW_RO model is a three-parameter type model adapted both for spiral-wound and hollow fibre type modules, which are currently the most common ones for industrial seawater desalination processes. SW stands for seawater, indicating that this model is to be used with seawater streams. The flow patterns of the different module types are currently not treated individually. The model very much follows the solution diffusion model which can be found in literature. With the following three parameters, module

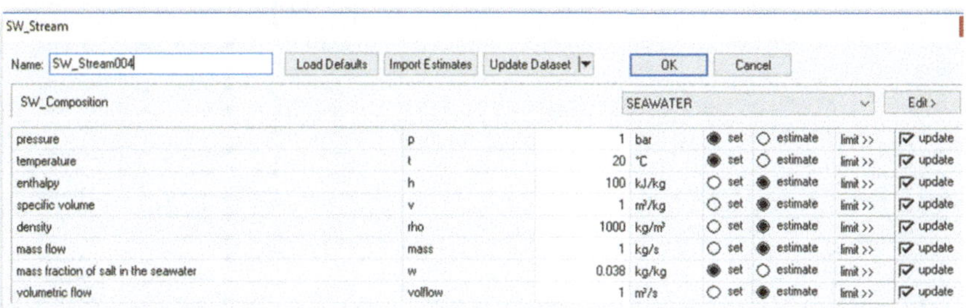

Figure 6.8 Standard SW_Stream data input dialog for sea water quality (MIDES Consortium).

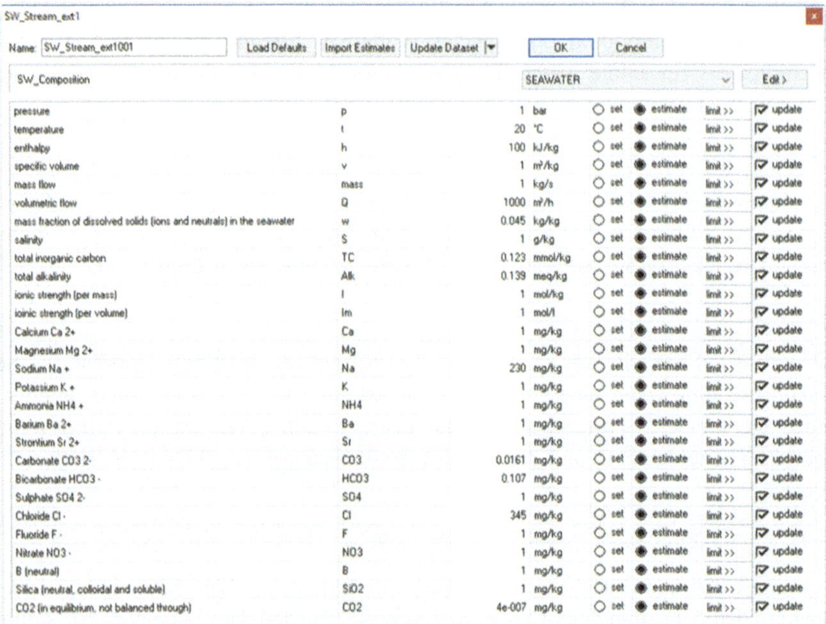

Figure 6.9 SW_Stream_ext1 data input dialog (MIDES Consortium).

performance is determined: Permeation coefficients A and B are membrane specific, the salt mass transfer coefficient k_s is dependent on geometry and flow conditions in the membrane module.

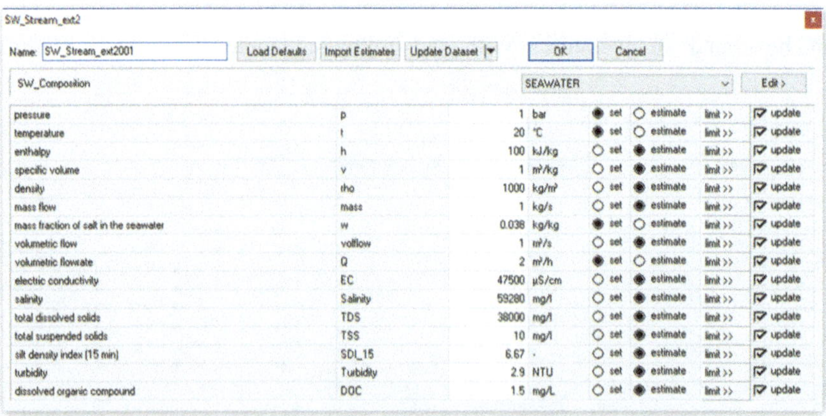

Figure 6.10 SW_Stream_ext2 data input dialog (MIDES Consortium).

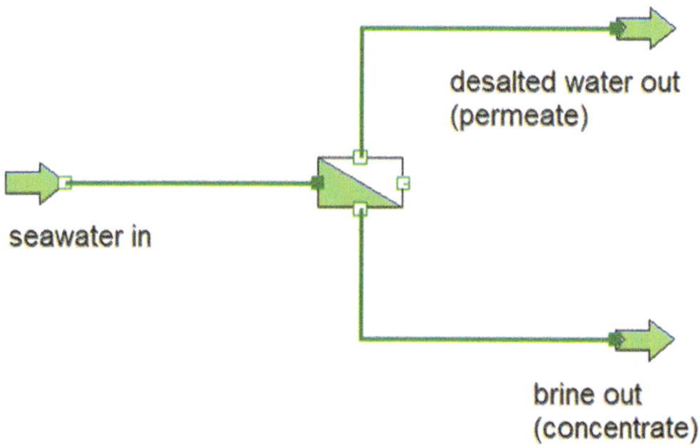

Figure 6.11 Single RO model implemented in IPSEpro (MIDES Consortium).

The individual RO model with input and output flows is displayed in Figure 6.11, followed by the data input dialog for the model (Figure 6.12).

Together with pumps and pressure recovery devices, arbitrary process configurations can be set up. Figures 6.13 and 6.14 display a typical RO configuration with energy recovery using a pressure exchanger and a two-stage RO process with tapered design. In both examples the single RO desalination model is used.

6.2.2.4 Electrical circuit models

For the calculation of electrochemical systems, like the MDC, the model library was extended by basic concepts of electrical circuits (transfer of current, local electric potentials, Kirchhoff laws, resistances). Several component models making use of these concepts were developed. As an example, the layout of an electrical circuit

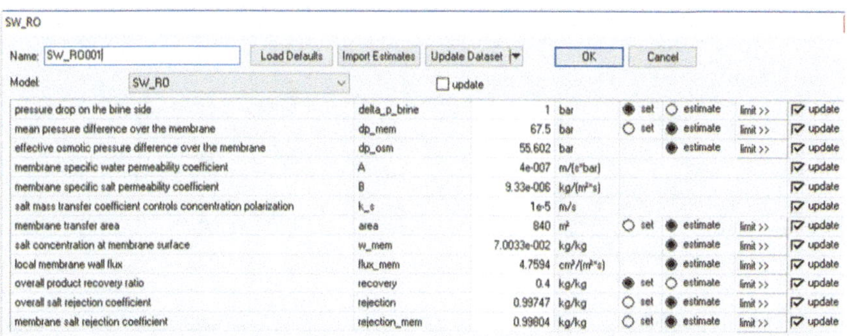

Figure 6.12 Data input dialog of an RO model (MIDES Consortium).

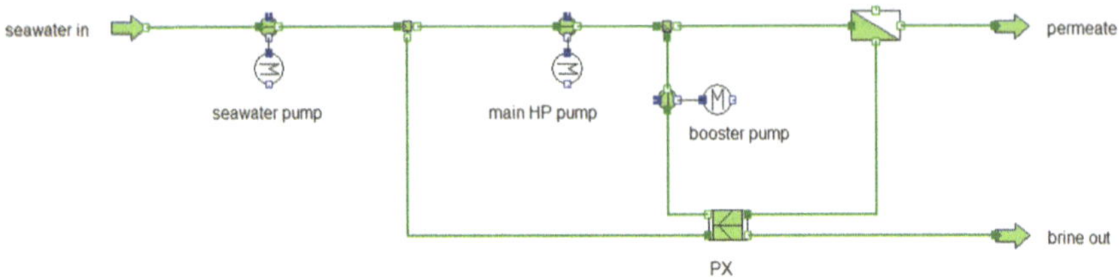

Figure 6.13 RO model with pressure recovery (PX: pressure exchanger for seawater) (MIDES Consortium).

with two general voltage sources in series and electric resistances in series and in parallel is shown in Figure 6.15. The values represented in the data-crosses are given according to the units of current (A) and voltage (V), as illustrated in the units data-cross at the bottom left part of the figure.

6.2.2.5 MDC models

The IPSEpro simulated models support understanding the performance of the involved processes. The MDC mathematical simulation model detailed in (MIDES Consortium; MIDES Consortium), was a useful means to predict the behaviour of MDC systems in different operating conditions, given by its parameter values. Hence, the mathematical process model was able to simulate accurately the quantitative influence of different parameters on the MDC performance. Figure 6.16 shows the first version of the Microbial Desalination Cell Model, as the core of the MIDES process in an IPSEpro model. It describes the process in one membrane stack based on published material, which claimed to present a mathematical model to simulate dynamic behaviour of MDCs for the first time through evaluating multiple factors such as organic supply, salt loading, and current generation.

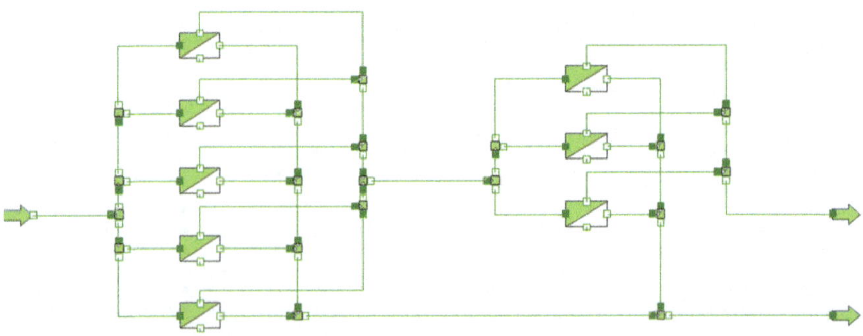

Figure 6.14 Two-stage RO process with tapered design (MIDES Consortium).

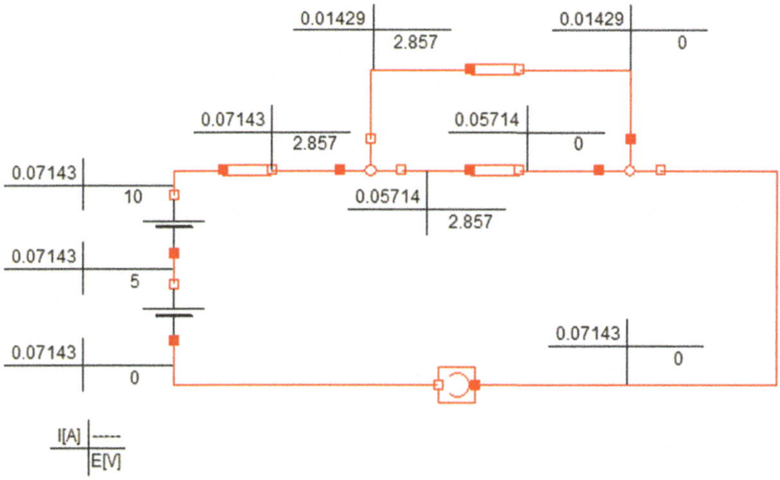

Figure 6.15 Electrical circuit model with two voltage sources and three resistances (MIDES Consortium).

The feeds are: one inlet for the salt water to be converted into diluate (product) water, and one inlet for the wastewater (anolyte) to become concentrated by the salt absorbed. The catholyte on the cathode side is not available explicitly. The drains are: one diluate (product) water stream and one concentrate wastewater stream with increased salinity. Connectors are available for the electrical circuit.

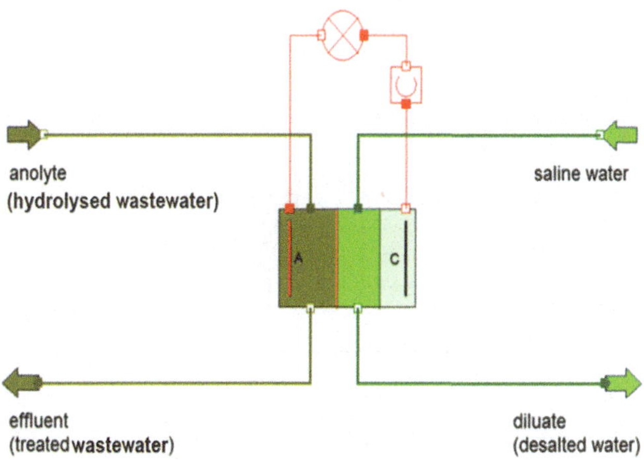

Figure 6.16 First MDC Model implemented in IPSEpro (MIDES Consortium).

6.2.2.5.1 MDC1_dyn_mod1 model

It is in parts based on the Microbial Desalination Cell Model 1 (Figure 6.16), but was modified and extended to better suit the needs of the MIDES application, as shown in Figure 6.17.

In the modified model the derivative of variable M_OX (i.e., the equation with M_OX') is disconnected from the other model equations. This modification allows decoupling of microbial activity from electrochemistry and gives more freedom to try out and test the remaining model. Of course, thereby, the anaerobic bacteria around the anode are not connected to the cell behaviour any more.

Tests with changes in input parameters and the influence on voltages, cell potential and resistances were run. The values displayed in this paragraph are only for demonstration of the current level of detail and were not matched or compared with the intended MIDES application.

In Figure 6.17, The values represented in the data-crosses are given according to the units illustrated in the two data-crosses of the bottom right part of the figure. The data input parameters for the implementation of the extended MDC model of Figure 6.17 can be seen in the IPSEpro dialog of Figure 6.18.

The complete MIDES process was implemented, simulated and optimized in the IPSEpro-PSE: Process Simulation Environment (SimTech Simulation Technology), as illustrated in Figure 6.19, including the MDC and the other customized desalination and wastewater-treatment components created using the IPSEpro-MDK (SimTech Simulation Technology) for the MIDES_Lib. This MIDES Process Model was validated by project partners, taking into consideration to the parameter values of the mathematical models and the measured data from the pre-pilot set-up.

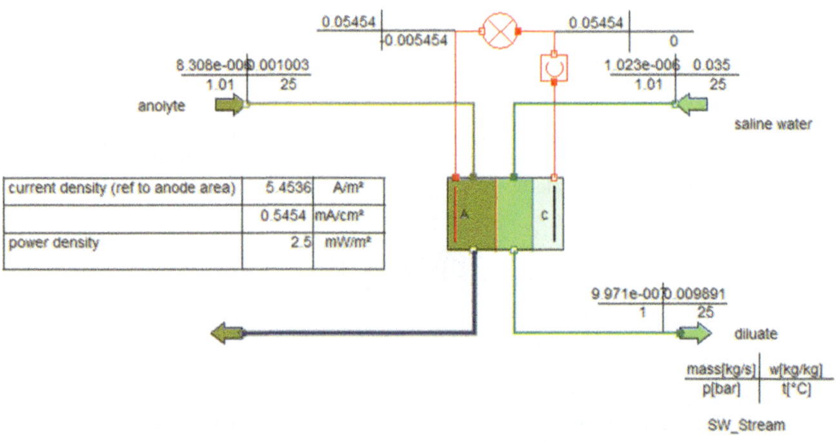

Figure 6.17 MDC1_dyn_mod1 extended model (MIDES Consortium).

Figure 6.18 Data input dialog of the extended MDC model (MIDES Consortium).

6.3 MIDES SOLUTION ENERGY CONSUMPTION & OPTIMIZATION ANALYSIS

This section presents some considerations regarding the energy consumption and gives an exemplary performance and optimization analysis of the overall

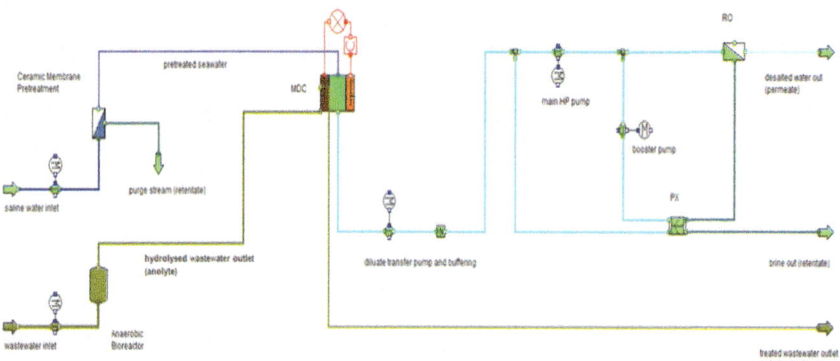

Figure 6.19 MIDES Process Model implemented in IPSEpro (MIDES Consortium).

MDC-MIDES solution. Energy plays an important role when looking at the performance process. This brief overview is from a deliverable of the MIDES project (MIDES Consortium), which also gives further details for comparison. As stated there, the specific energy consumption for water desalination (brackish water and seawater) lies within an average range of 0.53 to 5.3 kWh/m^3 within the water use cycle, with an estimated distribution of: 0 to 3.7 kWh/m^3 for water supply and conveyance; 0.03 to 4.2 kWh/m^3 for water treatment; 0.18 to 0.31 kWh/m^3 for water distribution; 0.11 to 0.32 kWh/m^3 for recycled water treatment & distribution; 0.26 to 1.2 kWh/m^3 for wastewater treatment & collection; and 0 to 0.11 kWh/m^3 for discharge.

The requirements for water supply and conveyance to the plant and subsequent delivery to end consumers (drinking water) or discharge (treated wastewater) are not included here in the comparison, as they are not directly related to the represented treatment processes.

For the performance and optimization analysis of the overall MDC-MIDES solution, several process models were developed within the IPSEpro environment (MIDES Consortium). In the following, a layout of the MIDES process at pre-pilot size is presented with a projected 18 L/day desalted water production. Brewery wastewater is used as the fuel feed source for the MDC. The saline water to be desalinated in the MDC and subsequent RO section is seawater with an inlet salinity of 35 g/L. Ceramic membrane pre-treatment of the seawater is installed. For reduced energy consumption, pressure exchangers are used in the RO section for modelling purposes.

All electrical power consumption is summarized and related to the drinking water outlet flow. Main contributors are the various pumps used throughout the process. At the moment no significant thermal power consumption is expected in the MIDES process. For the COD reduction in the MDC section, the model uses conversion factors which are hypothetical design values, as they could so far not be checked against measured data from a plant.

Figure 6.20 shows the calculation results of the MIDES overall model. Since more details for the model optimization will be available with the testing and start-up of the MIDES pilot plants, further feedback integration on this matter will be performed. IPSEpro simulation planned to consider the following two case studies: Demo Site 1: Brackish water stand-alone desalination using MDC; and Demo Site 2: Seawater desalination by MIDES technology (MDC-RO unit).

6.4 IPSEPRO ONLINE PLATFORM – IMPLEMENTATION

The ability to collaborate with others online has become a vital element of the modern workplace. Cloud-based systems for collaborative work on documents have made it much easier to develop projects with contributors in distributed locations. Cloud-based systems often do not require any software to be installed

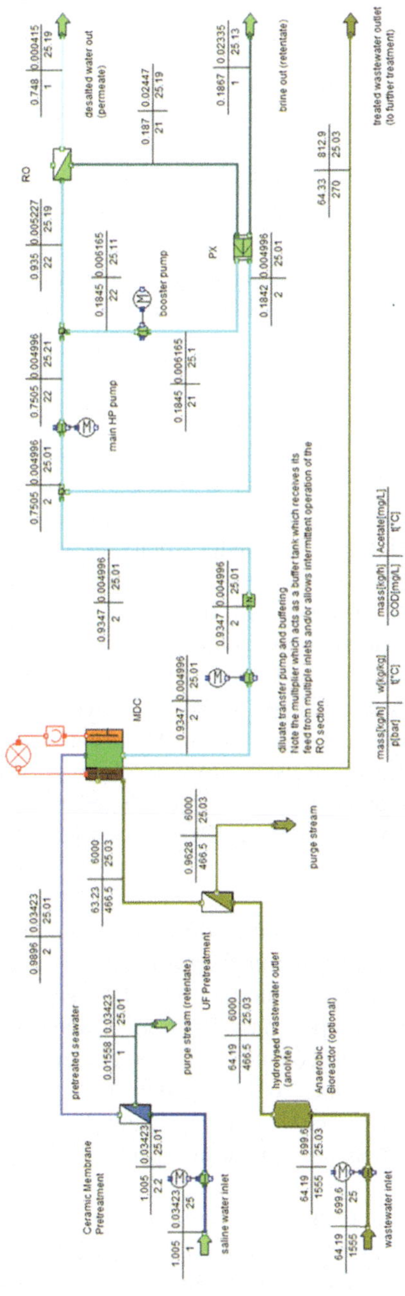

Figure 6.20 MIDES overall model calculation results (MIDES Consortium).

locally. Instead, a standard web browser functions as the user interface. From a user's perspective, using such systems generates several key benefits, such as:

- Since the documents are stored in the cloud and can be accessed from different locations, this enables collaborators to contribute to documents without the need to send copies to each other via e-mail or other communication means.
- Since a standard web browser serves as the user interface, the system is, to a large extent, hardware independent and can be used from a wide range of devices.
- It is not necessary to license and install software locally. It may be necessary to sign up for a service agreement, but typically this is more cost effective than the purchase of local software licences. It also ensures that any new functionality that becomes available can be used instantaneously by all users.

A requirement of the MIDES project was to present its overall process model in a shared online platform, to allow better collaboration among developers and project partners. By using the solver core of IPSEpro, it was possible to base the implementation of the cloud-based platform on a web modelling approach. Using the same solver core and the same model library as in the original IPSEpro environment ensures that the results obtained with the cloud-based platform are identical to those previously obtained with IPSEpro.

The work described hereafter concerns the development of the IPSEpro Online Platform (MIDES Consortium), with the characteristics described above for creating and solving process models, without the requirement to install any software locally. All interaction with the model, from defining the model to reporting results, is done via a web browser. Taking advantage of recent developments in browser technology, a browser-based user interface has been developed, as included in the general IPSEpro Online Platform concept in Figure 6.21; and in the system architecture at implementation-level, shown in Figure 6.22.

The IPSEpro cloud-based simulation platform has been implemented as a web application and as such typically defined as a client-server computer program, which the client (including the user interface and client-side logic) runs in a web browser. Consequently, the components of web applications can be divided into the categories: client-side components and server-side components, and a middleware which manages the communication between client and server. Figure 6.22 illustrates the overall implementation-level architecture of the IPSEpro cloud-based simulation platform, partially developed under the MIDES project.

Client-Side Components: In the cloud-based simulation platform, the client side comprises two major components, a project management component and the flowsheet editor. The project management component enables the user to log into the system, to create new projects, or select existing projects to use in the

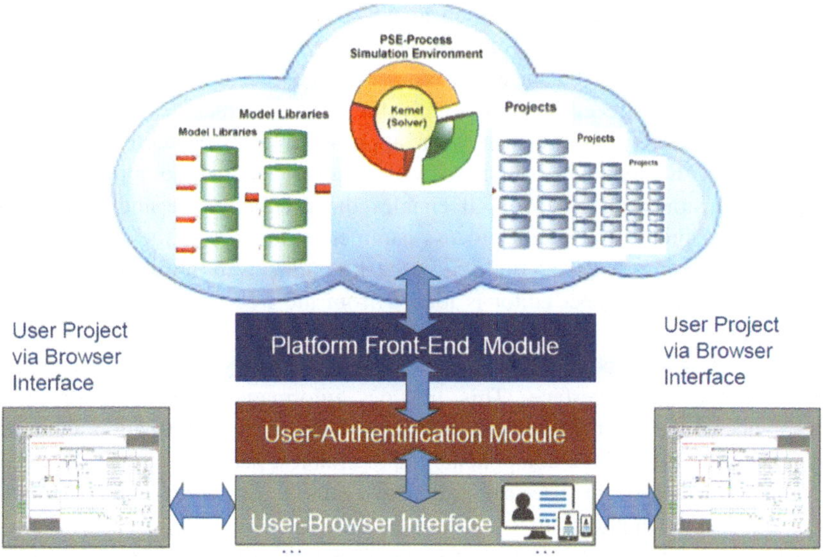

Figure 6.21 IPSEpro Online Platform concept (MIDES Consortium).

flowsheet editor. It also includes functionality to enable users to share projects with other users and to delete existing projects, in this way managing the data stored on the server. The flowsheet editor enables the user to create and modify the flowsheet of the process model, to edit process parameters, to execute the system solver and to display results.

Server-Side Components: The server-side of the cloud-based simulation platform includes two major components: (1) A web server with a database server, which stores all data about users and user projects. The web server also

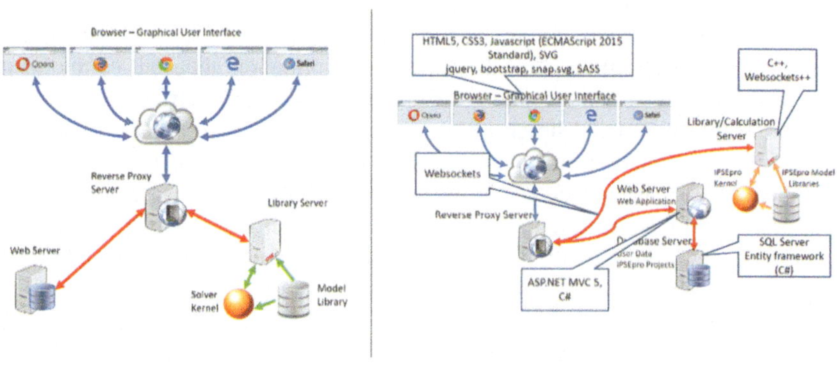

Figure 6.22 IPSEpro Online Platform - system architecture at implementation-level (MIDES Consortium).

provides all parts of the interface which are not specific to a particular model library; and (2) a library server, which is responsible for providing all model-library-specific information required by the client side and which provides the capability to solve a process model that it receives. Additionally, the server side includes a reverse proxy server which directs client requests to the appropriate back-end server.

Flowsheet Editor: The flowsheet editor is the central component of the cloud-based simulation platform. It enables the user to create and modify the flowsheets of processes, to edit process parameters, to execute the system solver and to display results. To edit a flowsheet, a new web browser window is opened, and the flowsheet editor is loaded from the web server. Once the basic functionality of the flowsheet editor is loaded, it requests from the library server all the information about components in a model library which is required to configure the user interface. This includes information about the graphical appearance of the available components (icons), as well as information about the user-accessible data in the components (i.e., variables, parameters, etc). After this information is loaded, the user can graphically edit the flowsheet in the same way as with a desktop application.

When the flowsheet is configured, the user can trigger the solution of the flowsheet. The flowsheet editor sends the request for solving the system to the library server. When it receives the results, it automatically displays them back in the flowsheet.

In order to ensure the long-term usability of the system on a wide range of platforms, it was decided to base the flowsheet editor on HTML5 (W3C, HTML5, 2014), which provides a rich set of capabilities that allows the system to implement interactive graphic functionality as required in the flowsheet editor. Moreover, the use of HTML5 allows the system to implement dynamic web pages without the need to use any browser add-ins, it is sufficiently well supported by all major web browsers (including Chrome, Edge, Firefox, Opera and Safari), and the IPSEpro cloud-based simulation platform has been successfully tested with all of them.

Library Server: The library server provides two services: (1) upon the respective request from a client, it returns the information about all components in a model library which is required to configure the user interface for using it with the respective model library; and (2) it also handles the request to solve a project. The information about the system components that are required by the flowsheet editor are extracted from the model library description used by IPSEpro. When the client requests to solve a process, it includes with the request a complete description of the structure and parameterization of the system. The library server feeds this information to the solver core, which converts it into a system of algebraic equations, which it then solves. The results are then sent back to the client to be displayed graphically.

Performance Aspects: The usability of a cloud-based platform is, to a large extent, determined by the system performance. Slow and unresponsive interaction will inevitably reduce acceptance by users. Several individual aspects have been evaluated during the development of the IPSEpro cloud-based simulation platform. They are:

- Time for opening a project/storing projects;
- Responsiveness of the user interface/browser; and
- Time required for solving a system.

Unlike with locally installed software, the performance of a cloud-based system depends on a wide range of factors, where the speed of local hardware is only of moderate impact. Some of these factors are: (i) the browser which is used on the client side; (ii) the speed of the internet connection; and (iii) the speed of the cloud server.

The time required for opening and storing a project is determined by the speed of the internet connection. The project shown in Figure 6.23 MIDES Process Model simulated in the IPSEpro Online Platform takes about 3.5 seconds to open. The responsiveness of the user interface is crucial for convenient use of the platform. Comparison between different hardware and software configurations show that the choice of web browser is a major influence, due to the highly interactive nature of the user interface.

The overall time required to obtain a solution for calculation of a project is determined by the speed of the internet connection for transferring input data to the library server and returning results back from the server, as well as the time required by the library server to actually solve the system of equations. The

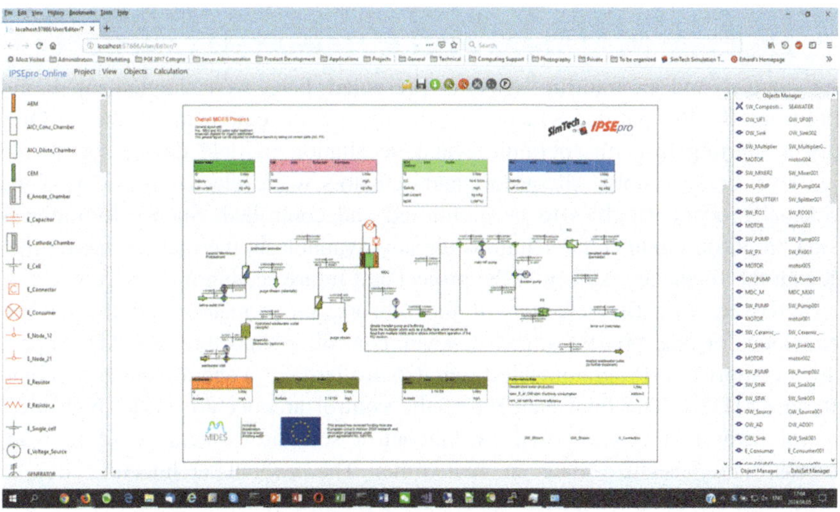

Figure 6.23 MIDES Process Model simulated in the IPSEpro Online Platform, shown as a screen-shot of a browser (MIDES Consortium).

solver core used by the library server is fast. The actual solution for the project shown in Figure 6.23, takes about 0.5 seconds. Measurements show that, on average, about twice this time is required to transfer data between the client and the server, so that the overall time for solving a system is about 1.5 seconds.

Platform Implementation Risks and Mitigation: Using a cloud-based system can present some risks due to the dependency on a public networked system. The system may have to cope with network failures, as well as with security threads. The implementation of the IPSEpro cloud-based simulation platform has addressed those risks and their mitigation measures. It is inherent in the system architecture that a network failure will result in severe restrictions on the use of the simulation platform. For example, in the case that it is impossible to reach the library server for obtaining simulation results. However, the implemented functionality of the platform ensures that even in the case of a network failure, no data is lost. Likewise, security procedures have been implemented to protect user data from unauthorized access.

6.4.1 MIDES desalination process model used online

Figure 6.23 shows the MIDES Process Model within the IPSEpro Online Platform (MIDES Consortium). It is possible to identify on the left side of the environment, the icons of the customized desalination and wastewater-treatment components created for the MIDES_Lib model library and used in the simulation of the overall MIDES process. Results of the simulation are automatically displayed in the data-crosses placed along the process components' connections; and within the parameter boxes in the flowsheet.

6.5 MIDES MONITORING AND AUTOMATED CONTROL

The control and monitoring systems presented here were implemented and optimized by the partner, Oncontrol, during the development of the MIDES project, starting from the definition, and installation of field devices up to the high-level control tools. All sensors and actuators were selected considering the principal process variables to be monitored and controlled. Such selection was made based on commercial solutions already available in the market, taking into account the financial viability of the project and future scaling up.

Moreover, with the selection of the sensors, the actuators and the desired behaviour for the process, the low-level control architecture was developed, which comprises a Programmable Logic Controller (PLC) and a Human Machine Interface (HMI). The final element of the control architecture is the high-level control layer that is composed by Oncontrol software systems used to read, process, and visualize data in real time. The overall architecture of the implemented system is depicted in Figure 6.24.

As illustrated in Figure 6.24, the data module Databridge is responsible for all data acquisition, processing and data storage. Databridge is a software solution

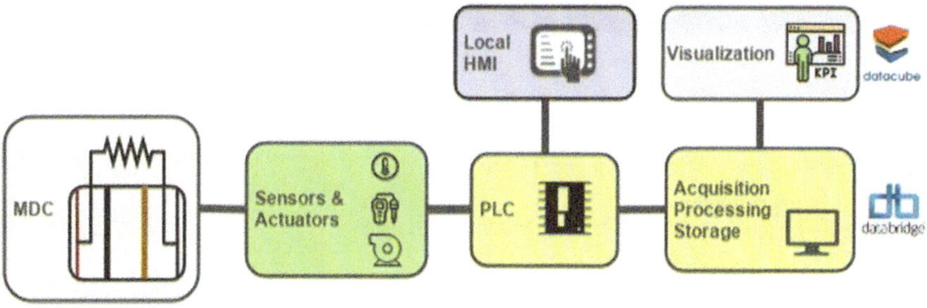

Figure 6.24 Control system architecture (MIDES Consortium).

from Oncontrol that executes an extract-transform-load (ETL) pipeline for industrial data processing. It is modular and has a user-friendly interface. In short, Databridge has built-in modules to execute machine learning, fuzzy logic, fuzzy, predictive controllers, and customized scripts. It processes data from different data sources (Extract), such as OPC, OPC-UA, CSV, Modbus. Afterwards, the data is processed in the built-in modules (Transform), loading back the results to any input source (Load), for example to a PLC, or to a relational historical database, saving it in a PostgreSQL database, or CSV file.

For the MIDES project, the main Databridge plugins used were the following: (1) Expert Fuzzy Control System: software that allows the incorporation of expert human knowledge about the process in the system to be controlled using "if-then" fuzzy rules. The user can construct the controller and adapt it using a user-friendly graphical interface; (2) Statistical Modeling System; (3) Ontrend Plugin: software that allows the creation of historical data of the process, saving it in a PostgreSQL database; and (4) PostgreSQL: Database used by the Ontrend plugin of Datacube to show/monitor the historical data of the process.

In addition to Databridge, Datacube was also used to achieve unattended control architecture and process monitoring. Datacube is a high-level user interface platform that is used for: historical analysis of the process using the plugin Ontrend; as well as for analysis of the key performance indicators of the process. The Datacube system is a modular and extensible software, which creates an interface between the operator and the process. It allows for a historical view of the process, online, and automatic creation of reports, and visualization of key performance indicators (KPIs) of the process in a dashboard presentation style. To monitor the performance of the MIDES process, a tool for the implementation and monitorization of the KPIs of the process was created. Using this platform, the most varied KPIs can be implemented: overall process efficiency, production flow, and energy management system, among others. As can be seen in Figure 6.25, the software offers a graphical interface using Datacube that allows the user to monitor the KPIs and check historical data. Such a tool is particularly important to support decision-making processes within the project implementation procedures.

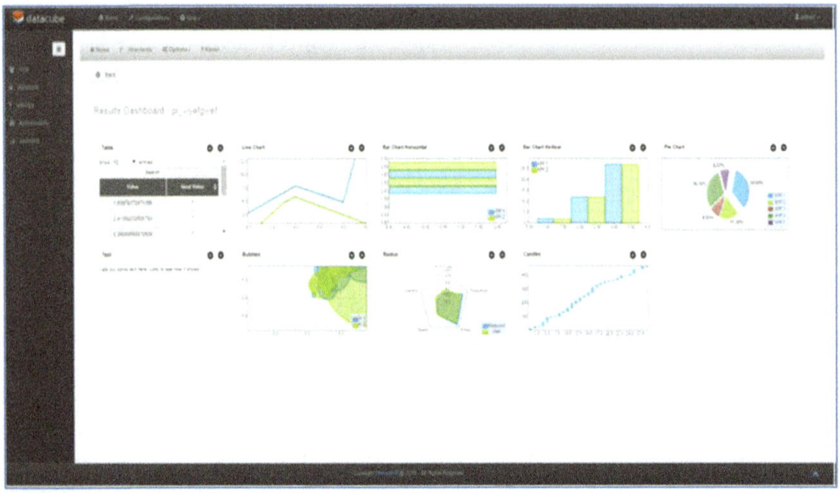

Figure 6.25 Datacube key performance indicators (MIDES Consortium).

6.5.1 Monitoring control system in the pre-pilot

In order to design the control and monitoring systems, the following steps were accomplished in MIDES:

(1) Design of instrumentation. In this step the instrumentation needed to control and monitor the process (i.e., current, pH, conductivity, temperature) were chosen from available equipment from major industrial equipment manufacturers.
(2) Selection of the actuators for the process. In this step, taking in account the needs of the process, the valves and pumps to control the process were selected.

At the end of these two steps, taking into account the selected sensors and actuators, the piping and instrumentation diagram (P&ID) was attained. For the process monitoring, the following key variables were defined to be monitored:

- pH of the anolyte, catholyte and saline tanks;
- Temperature of the anolyte, catholyte and saline tanks;
- Redox value of the anolyte, catholyte and saline tanks;
- Conductivity of the anolyte, catholyte and saline tanks;
- Anolyte potential;
- Catholyte potential;
- Potential of the anode in each MDC;
- Current that flows in each MDC.

In addition to those sensors, the analogue inputs of the PLC were used to measure the following remaining variables of the process:

- Potential of the anode in each MDC;
- Potential of the anode and cathode feeding using electrode sensors that are in the feeding pipes.
- Current that flows in each MDC, measuring the potential in the terminal of low resistance/shunt with a well-known resistance value. When a current passes through the shunt, a proportional millivolt output is produced.

As previously mentioned, to read data from the sensors and also to interact with the actuators, a Programmable Logic Controller (PLC) was used. The PLC is a crucial part of the control of the MIDES pre-pilot. It receives the desired behaviour of the process, which the operator configures in the HMI, and controls the actuators, based on feedback from the sensors, to achieve the desired behaviour.

The next step taken was the design of the electrical diagram for the construction of the switchboard. From this electrical diagram, the switchboard was assembled. Overall, the main characteristics of the switchboard are: IP66 wall-mounted polyester enclosure, safety equipment for devices and personnel protection, power supply of the automation equipment, sensor and actuators power supply of the MDC; and automation equipment: PLC and HMI. Figure 6.26 shows the external and internal view of pre-pilot control system switchboard.

In parallel to the switchboard construction, the programming of the PLC and HMI was performed (Figure 6.27). As previously mentioned, the user configures

Figure 6.26 MIDES pre-pilot control system switchboard: external and internal views (MIDES Consortium).

(a)

(b)

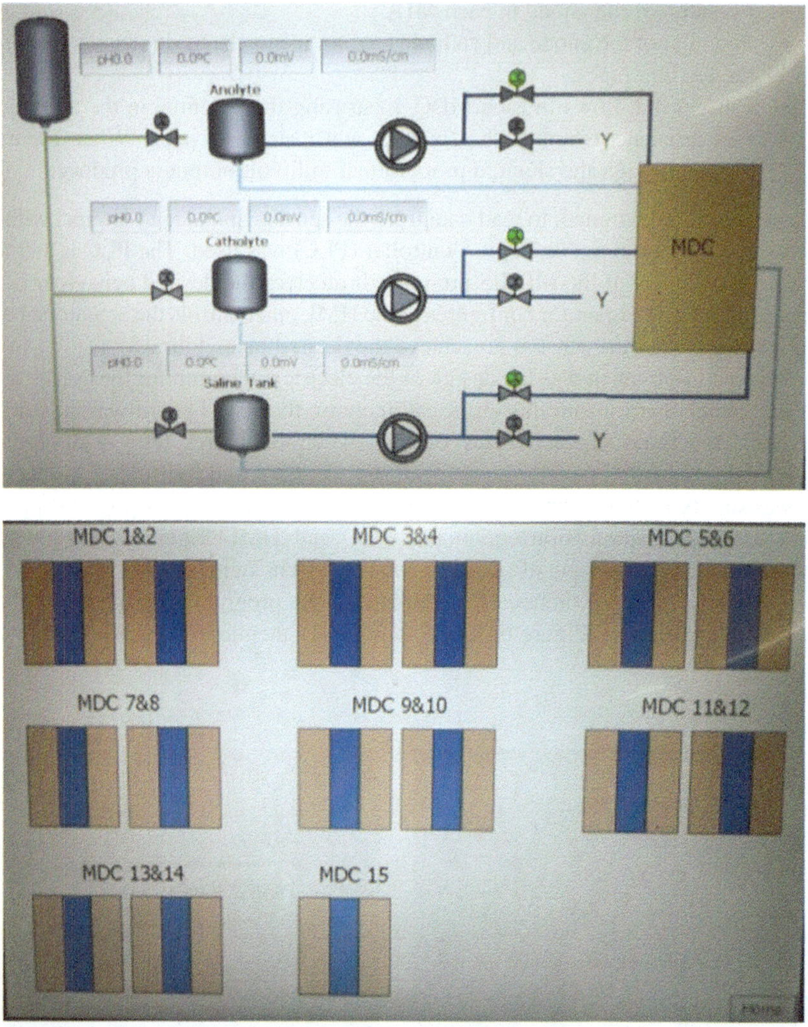

Figure 6.27 HMI interface: (a) pre-pilot general control; (b) MDC individual control (MIDES Consortium).

the desired behaviour for the process via the HMI and the PLC controls the actuator to achieve the desired setpoints, having feedback from the installed sensors.

As can be seen in Figure 6.27, in the visual interface of the pre-pilot, the user can control the peristaltic pumps, the liquid solenoid valves and the gas solenoid valves. Furthermore, the user can also select the MDC cell to be configured and in its individual display the user can configure if the MDC cell is in polarization or not, monitor the different potentials of feeding and MDC, and also monitor the current of the MDC.

The control switchboard was installed in the MDC process in the IMDEA Water facilities. Jointly with the switchboard, all the sensors, actuators, and cables were also installed. Figure 6.28 shows the MIDES pre-pilot with one MDC in the stack.

Figure 6.29 parts (a), (b), (c) and (d) show more details of the MIDES pre-pilot installation.

6.5.2 Machine learning using databridge

As a sub-area of artificial intelligence (AI), machine learning (ML) focuses on studying algorithms and statistical models used by computer systems to perform a specific task, relying on patterns and inferences. ML algorithms build a mathematical model based on sample data (training data), in order to make predictions or decisions without using explicit programmed instructions (Bishop, 2006). ML can be applied to various applications across business problems, together with data mining (Friedman, 1998), focusing on exploratory data analysis for predictive analytics.

In MIDES, an ML approach with statistical implementation was developed using the Databridge model to predict the chemical oxygen demand (COD), by means of a

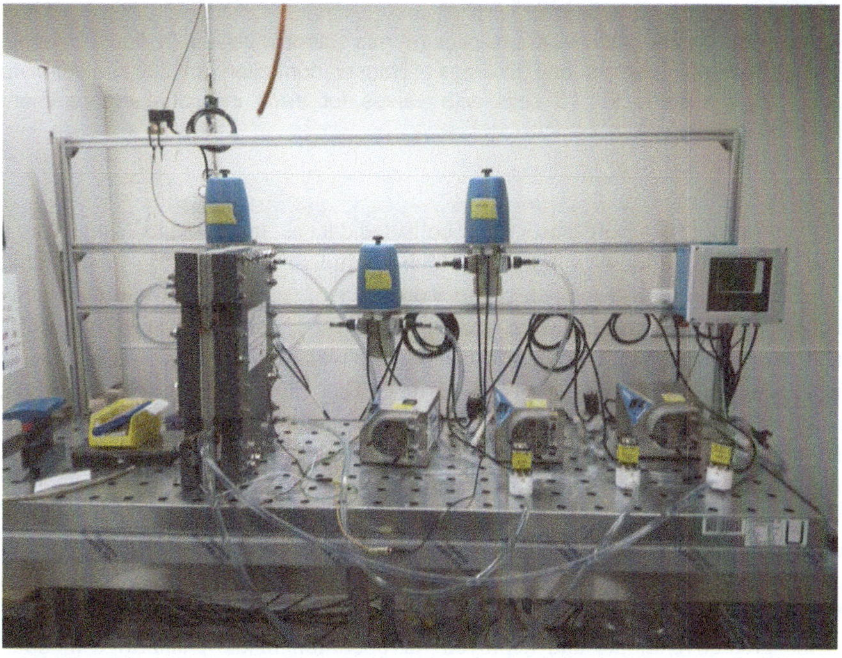

Figure 6.28 Overview of the MIDES pre-pilot (MIDES Consortium).

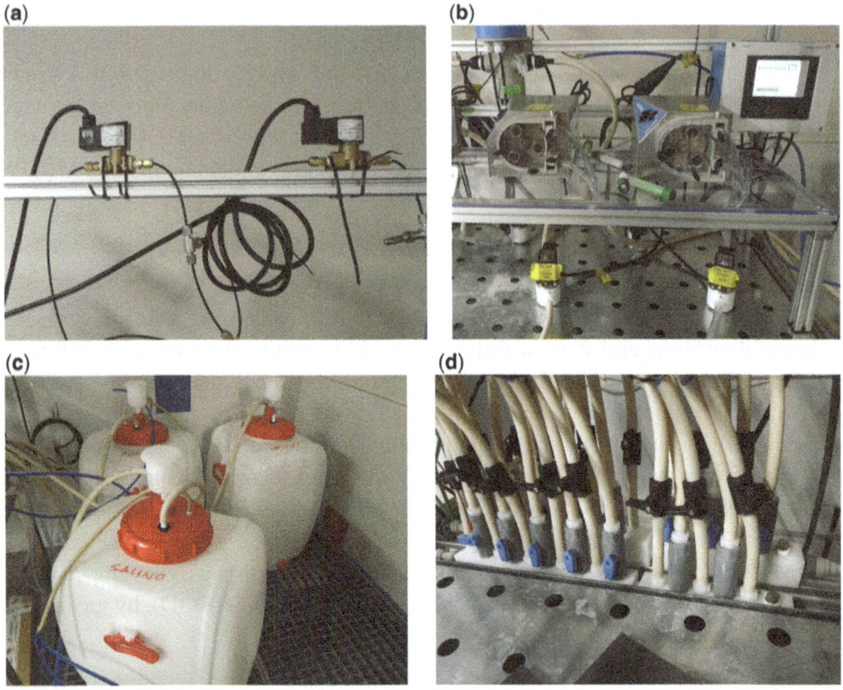

Figure 6.29 Details of the MDC pre-pilot: (a) gas solenoid valves; (b) liquid solenoid valves, peristaltic pumps, and Endress +Hauser controller; (c) 200 L electrolyte tanks; (d) manifolds for 15 cells with valves for each individual compartment (MIDES Consortium).

software sensor (aka. soft sensor). The collected data as 'training data' came from a total of 12 monitored variables, which included: conductivity, pH, redox, temperature, current generated, pump velocity, etc. (in catholyte, anolyte, saline tanks). Those parameters were measured every 1 minute. Data from COD laboratory measurements were also considered, which have scarce and infrequent measurements, approximately twice a day. The COD soft sensor allows the estimation of COD in real time at every 1 minute, allowing the operators to check the trend of COD, and take corrective actions, such as adding more organic matter to the anolyte solution to accelerate the desalination rate.

To predict the COD, the variable parameters from saline, anolyte and catholyte tanks were used as input variables. Those were: the conductivity, pH, redox and temperature, for which a total of 16 samples were used to learn the COD predictive model. For such purpose, a partial least squares (PLS) model was used as the predictive model and deployed for operation in the Databridge module. The PLS was chosen as the predictive model, because it is robust to noise, and correlated and irrelevant features, which are common issues in industrial applications.

Figure 6.30 shows the coefficients of the PLS model, and the prediction comparison between real and estimated COD. Figure 6.30(a) shows the relevance of each input variable regarding the COD values. It can be concluded that conductivity in the anode has the strongest correlation with respect to COD (which is in line with MDC operating conditions, since the conductivity is related with ionic conductivity of organic matter in the anode tank), along with the saline redox (in MDC technology, the redox values tend to positive values when oxidizing species appear in the solution. For example, oxygen or Fe(III) that can be reduced into water and oxidize other molecules). Figure 6.30(b) shows the coefficients of the PLS model for COD prediction (red), it can be noticed that there is a pertinent and well-accepted agreement between prediction and real values, which validates the ML approach.

6.6 OPERATIONAL DECISION MAKING SUPPORT FOR THE MIDES PILOT PLANT

Operational decision support systems have been extensively studied, including use-cases (InteGRail FP6 Research Project, 2008; Nurminen *et al.*, 2008), and have been shown to offer decision-makers accurate and appropriate conditions for better decisions, as they are provided with the required information and results from simulated models of their analysed processes, in order optimize operations as well as further implementation phases of respective systems and installation plants.

Within MIDES, the operational decision making support provided by the simulation analysis of the MDC and its process, as well as its automated operational control was of vital importance for the accuracy of its performance

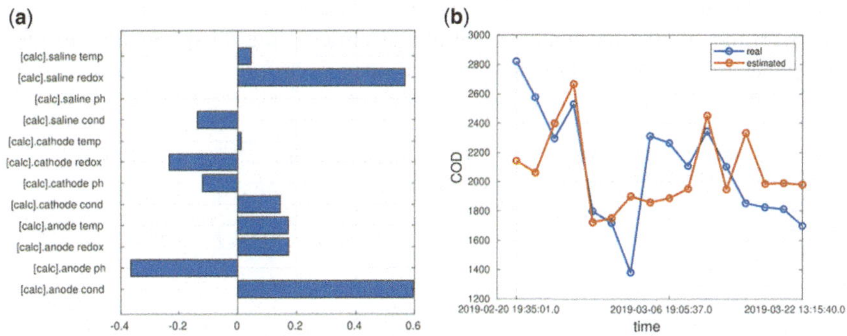

Figure 6.30 (a) Coefficients of PLS model for COD prediction, (b) Predicted and real COD (MIDES Consortium).

and data validation, as well as for the mitigation of risks in the pilot implementation stages.

The mathematical model of the MDC provided by the results of project work-package concerning '*Microbial Cell Design Engineering and Testing*' were further implemented in IPSEpro and integrated in the customized MIDES component-model library (MIDES_Lib), which augments the existing IPSEpro desalination model library (SimTech Simulation Technology). The MIDES_Lib was then used in the IPSEpro Process Simulation Environment (PSE) to create the comprehensive MIDES Process Model, which was simulated, optimized and validated by all involved technical partners.

The concept of using IPSEpro as a decision support system (DSS) or coupled with a DSS for operational applications has its origin described in (Dargam & Perz, 1998) and has been a source of insights for further investigations in the DSS area (such as in Kazim & Aydin, 2011, for instance).

Specifically, the analysis planned to support operations in MIDES was based on the insights and results of the respective simulated process model, as well as of its monitored control system. Figure 6.31 shows the considered steps used in MIDES for operational decision support, in order to allow pertinent actions to be taken to improve system performance envisaging more accuracy in the implementation phases of the project. Steps 1 to 6 were planned to be taken using the MIDES mathematical model and the simulated process model in IPSEpro. Step 7 relates to the actions implemented in the pilots implementation phase. Finally, steps 8 and 9 relate to the performance of the automated control & monitoring system built within the pilots. A feedback from step 9 to step 4 was needed in some cases to re-start the process with updated values.

The overall simulated model implemented via SimTech's process modelling package IPSEpro, including its developed and deployed online platform, was then a useful asset for the project, in providing a decision-making tool concerning the integration of the MIDES process from lab-scale to pre-pilot scale. The use of the simulated model of MIDES allowed confirmation of expected parameters and values of the project implementation that then validated its implementation phase. Moreover, the simulation model of the overall process provided valuable insights at all stages of the project (lab, pre-pilot and pilot scales).

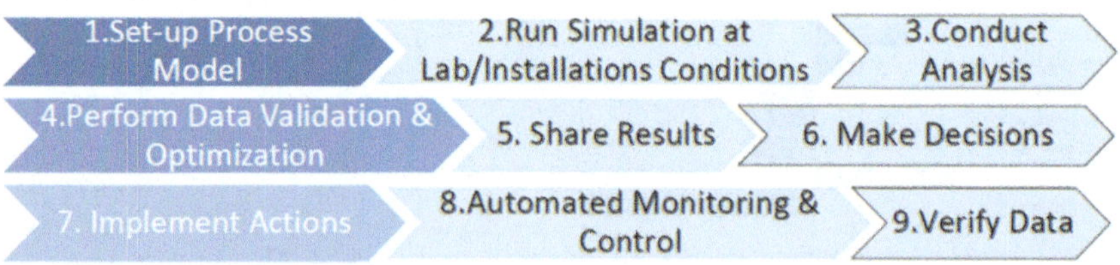

Figure 6.31 Steps of the analysis planned for operational decision support in MIDES (Dargam *et al.*, 2020).

Furthermore, the results obtained in the analysis of the MIDES pre-pilot-scale MDC could also be used for assessing the impact of up-scaling of changed component parameters in the next phase of the pilot-scale MDC, and even analysing the performance of the pilot plants.

Likewise, the results of the MIDES monitoring system allows project managers, end-users and decision-makers to have a clear picture of the operational situation of the pilot plants in order to be granted assistance on assertive potential decision-making options, whenever needed.

6.7 SUMMARY & CONCLUSIONS

In conclusion, the outcomes concerning the modelling and simulation as well as the performance analysis and optimization of the MIDES overall process, were presented. From the individual component models that compose MDC, together with the pre- and post-treatment component models, it was possible to set up the preliminary overall process model for the microbial desalination cell solution in order to study and perform an optimization analysis of its energy consumption and overall condition. A dedicated IPSEpro library of components for MIDES (MIDES_Lib) was developed, including existing components of the IPSEpro desalination library (Desal_Lib), various components needed in the overall MIDES process mainly for pre- and post-treatment, and several component models specifically developed for the microbial desalination cell solution. A series of process models were set up in IPSEpro, resulting from the simulation of the MDC process and from its performance analysis and optimization.

The MDC mathematical simulation models defined and implemented compose a useful tool to predict the behaviour of MDC systems in different conditions. They allow evaluation of the influence of different parameters on the production of desalinated water.

The presented work confirmed the claim that the simulated models can support understanding of the performance of the processes involved. This way, insights coming from the simulated process models can be used to confirm hypotheses assumed in the project proposal, as well as to recognize new behaviour of the MIDES technology in relation to process efficiency improvements, and reduction of electric energy requirements.

A cloud-based simulation platform for the IPSEpro process and simulation modelling environment was implemented and its capabilities were demonstrated. The experience gained with the system shows that it is well suited for realistic applications and can be particularly useful for collaborative project development, like MIDES.

IPSEpro Online Platform: a major benefit of the cloud-based simulation platform is the fact that no software needs to be installed locally. This makes the system access easier, which is of particular benefit in collaborative research projects where a larger number of users need to access the models. Additionally, there is no risk of version conflicts that can occur with locally installed software.

Considering the implementation aspects of the IPSEpro Online Platform and the Control Monitoring System, we can assure that the shared MIDES Process Model for development collaboration among project partners is robust and accurate in performance and security aspects; and that the automatically monitored data presented to the end-users and pilot developers, including machine learning and statistical analysis, serve as important tools of decision-making with accurate levels of assertiveness. Those features enforce a rapid decision-making process for potential changes to be implemented in the model as a whole, impacting all stages of the project implementation from lab-scale, through pre-pilot to pilot stages.

REFERENCES

Aneke M., Agnew B. and Underwood C. (2011). Performance analysis of the chena binary geothermal power plant. *Applied Thermal Engineering*.

Arevalo. (2018) Selecting Fuel Sources for Simultaneous Wastewater Reuse and Water Desalination Using Bioelectrochemical Systems. Presentation at the IDA Conference on Water Reuse and Recycling, Valencia, Spain.

Bishop C. M. (2006) Pattern Recognition and Machine Learning. Springer, ISBN 978–0-387-31073-2.

Dargam F. C. C. and Perz E. (1998). A Decision Support System for Power Plant Design. European Journal of Operational Research – EJOR 109, Special Issue on Decision Support Systems, pp. 310–320, Elsevier Science.

Dargam F., Perz E., Bergmann S., Rodionova E., Sousa P., Souza F. A. A., Matias T., Ortiz J. M., Esteve-Nuñez A., Rodenas P. and Zamora P. (2020). 'Supporting operational decisions on desalination plants from process modelling and simulation to monitoring and automated control with machine learning'. In: Springer Series LNBIP 384 – Lecture Notes in Business Information Processing (LNBIP_DSS) on 'Decision Support Systems X – Cognitive Decision Support System and Technologies', J. M. Moreno-Jiménez, I. Linden, F. Dargam and U Jayawickrama (eds.), ISBN: 978-3-030-46223-9, Springer Nature, Switzerland. doi: 10.1007/978-3-030-46224-6

Friedman J. H. (1998) Data Mining and Statistics: What's the connection? *Computing Science and Statistics*, **29**(1), 3–9.

Horizon 2020 Project MIDES. Horizon 2020 Project MIDES https://midesh2020.eu/, MIDES – Microbial Desalination for Low Energy Drinking Water. DoA – Description of Action (H2020-NMP-2014-2015/H2020-NMP-2015-two-stage). In European Commission – Horizon 2020 Project Grant Agreement No. 685793, pp 83–231. Associated with document Ref. Ares (2016)1046927, 2016 (Accessed 1 March 2021).

InteGRail FP6 Research Project. InteGRail FP6 Research Project, Intelligent Integration of Railway Systems, (2005–2008). http://www.integrail.eu/documents/fs21.pdf.Semiat R (2008) Energy Issues in Desalination Processes. Environmental Science & Technology 42: 8193-8201, doi: 10.1021/es801330u (Accessed 15 November 2020).

Karellas S., Leontaritis A.-D., Panousis G., Bellos E. and Kakaras E. (2012). Energetic and ex-ergetic analysis of waste heat recovery systems in the cement industry, Proceedings of ECOS 2012. The 25th International Conference on Efficiency, Cost, Optimization, Sim-ulation and Environmental Impact of Energy Systems, Perugia, Italy.

Kazim B. A. and Aydin U. (2011). A Multiple Criteria Energy Decision Support System, Techno-logical and economic development of Economy, ISSN 2029-4913 print/ISSN 2029-4921 online, 2011 Volume **17**(2), 219–245. Copyright © 2011 Vilnius Gediminas Technical University Press Technika, doi: 10.3846/20294913.2011.580563, 2011.

MIDES Consortium. MIDES Consortium, Deliverable D3.1 Definition of the MDC mathematical model for the lab scale set-up operation. MIDES Project – Published in 2016 under the Horizon 2020 Project Grant Agreement No. 685793.

MIDES Consortium. MIDES Consortium, Deliverable D5.1 Comparison Study of Energy Performance for var-ious Desalination Technologies. Published in 2017 under the MIDES Project – Horizon 2020 Project Grant Agreement No. 685793.

MIDES Consortium. MIDES Consortium, Deliverable D5.2 IPSEpro Model Library & Simulation of MIDES Process and Energy Consumption Study & Optimisation Analysis. Published in 2018 un-der the MIDES Project – Horizon 2020 Project Grant Agreement No. 685793.

MIDES Consortium. MIDES Consortium, Deliverable D5.3 – IPSEpro Models for MIDES Process & MIDES Web-based Process Model. Published in 2018 under the MIDES Project – Horizon 2020 Project Grant Agreement No. 685793.

MIDES Consortium. MIDES Consortium, Deliverable D5.4 – Overall MIDES automation project and Key Per-formance Evaluation. Published in 2018 under the MIDES Project – Horizon 2020 Project Grant Agreement No. 685793.

Nurminen M., Suominen P., Äyrämö S. and Kärkkäinen T. (2008). Use cases for operational de-cision support system. *VTT Research Notes*, **2442**, 107–131 +app 5p. Normative ver-sion: http://www.vtt.fi/inf/pdf/tiedotteet/2008/T2442.pdf (Accessed 01 February 2021).

Ortiz J., Ramírez-Moreno M., Zamora P., Arévalo J., Monsalvo V., Rogalla F. and Esteve-Nuñez A. (2018). Microbial Desalination Cell for brackish water desalination powered by energy contained in wastewater. Poster presentation at the 15th IWA-LET Conference, Nanjing, 2018.

Perz E. and Erbes M. (2011). Process Modelling of Organic Rankine Cycles. ORC, First In-ternational Seminar on ORC Power Systems, TU Delft, 22-23.9.2011, 2011.

Perz E., Riesel U. and Schinagl H. A. (1995). A new approach for modelling energy systems. ASME Cogen Turbo Power Conference, Vienna/Austria, August 23 – 25.

SimTech Simulation Technology. SimTech Simulation Technology, IPSEpro-PSE: Process Simulation Environment – User Manual. SimTech GmbH www.simtechnology.com (Accessed 15 November 2020).

SimTech Simulation Technology. SimTech Simulation Technology, IPSEpro-MDK: Model Development Kit – User Manu-al. SimTech GmbH www.simtechnology.com (Accessed 15 November 2020).

SimTech Simulation Technology. SimTech Simulation Technology, IPSEpro: Specification of the Desalination Library (Desal_Lib). SimTech GmbH www.simtechnology.com (Accessed 15 November 2020).

W3C, HTML5. W3C, HTML5: A vocabulary and associated APIs for HTML and XHTML. World Wide Web Consortium, W3C Recommendation 28 October 2014. https://html.spec.whatwg.org/ (Accessed 11 April 2021).

Zamora P., Asensio Y., Ortiz J., Esteve-Núñez A., Arévalo J., Rogalla F. and Monsalvo V. (2018). Anaerobic wastewater pretreatment as fuel production stage for Microbial Desalination Cells. Poster presentation at the 15th IWA-LET Conference, Nanjing, China.

Zamora P., Ramírez-Moreno M. and Ortiz J. M. (2019). Towards the World's Largest Micro-bial Desalination Cell for Low Energy Drinking Water Production. Proceedings of the IDAWC19 International Desalination Association World Congress on Desalination and Water Reuse 2019/Dubai, UAE, 2019.

Chapter 7

Environmental assessment for desalination projects including microbial desalination cells

Angeles Mendoza-Sammet[1], Sonam Jamtsho[1], Sergio G. Salinas-Rodríguez[1], Maria D. Kennedy[1], Nathan Bossa[2], Camilla Delpivo[2], Maria Diez[2] and Verónica González Andrés[2]

[1]IHE Delft Institute for Water Education, Delft, Netherlands
[2]Leitat Technology Centre, Barcelona, Spain

ABSTRACT

For MIDES it was necessary to analyse the aspects that should be considered for the environmental assessment of desalination projects that integrate the microbial desalination cells as pre-treatment for reverse osmosis. The approach used to that end was to combine three methods: revision of existing guidance for environmental and social assessment of desalination projects to compile a list of components and impacts; life cycle analysis based on a pre-pilot MDC to determine impacts associated with its construction and operation; and risk assessment to determine risks of the potential release of nanoparticles. The results showed that, at project level, existing guidance needs to pay more attention to the effects that the local and regional environment could have on a desalination project and to the regional cumulative effects to which a project could contribute. At the level of the MDC, the main impacts were the contribution to climate change and toxicity resulting from the 20:1 ratio of catholyte to feed water; however, preliminary data for the operation of a pilot MDC showed a reduction to a 5:1

doi: 10.2166/9781789062120_0175

ratio, with largely reduced impacts. At the structural level, the risk assessment showed that the release of nanoparticles from the cathodes and membranes is negligible.

Keywords: Desalination, Global Water Intelligence (2020): Environmental and Social Impact Assessment, life cycle assessment, MDC, microbial desalination cells, nanoparticles, risk assessment, technologies.

7.1 INTRODUCTION

Desalination has been known for centuries as a process to make seawater drinkable. The need to find ways to use saline water for drinking is part of human history. The bible includes interventions to make salty water drinkable and distillation has been reported as a method to desalinate water since at least the 900s. By the end of the 1500s experimentation was ongoing to build devices that could desalinate seawater (Nebbia & Menozzi, 1966). However, the commercial use of desalination plants started around 1869 in England to supply drinking water for ships (Lerner, 2020). In 1907, the first desalination plants were built in Saudi Arabia (Al-Mutaz, 1991) and in 1965 the first reverse osmosis desalination plant was built in California, USA (Apec Water, 2017). Since then, the number of desalination plants has increased to around 800 plants by 1974 (Shipman, 1974), 4,600 plants by 1985 (Al-Mutaz, 1991), and 20,956 by 2019, corresponding to a global installed capacity to produce close to 115.5 million m^3/d (Global Water Intelligence, 2020). As presented in the first chapter of this book, according to Global Water Intelligence (2020), most of the installed desalination capacity in million m^3 is in Saudi Arabia (close to 19), United States of America (13.7), United Arab Emirates (11.5), China (9.7), and Spain (5.8). The use of desalination is expected to increase as water scarcity also increases because of the increase in population, the deficient management of water resources and the changes in precipitation related to climate change (e.g., Mekonnen & Hoekstra, 2016). With that, the impacts of desalination may increase too.

In contrast to the interest in desalination, the need to prevent the impacts of human activities on the environment did not start until 1969–70, when the Environmental Protection Act, the Clean Water Act and other environmental laws were enacted in the USA. By 1972, only the United States, Norway and Sweden had enacted environmental laws, and by 1992 only 76 countries out of 179 had adopted an environmental framework. By 2017 the number of countries and other states with environmental frameworks had increased to 187, plus 11 more that have legal instruments that require an environmental assessment (United Nations Environment Programme [UNEP], 2019). As a reference, in 2019 the United Nations (UN) included 250 countries in its list.

Up to the 1970s, the discussions on desalination were related to its economic feasibility and the technological alternatives. For example, Shipman (1974) wrote a report on water desalination for the Public Utilities Department of the International World Bank for Reconstruction and Development. The report elaborated on cost, water pricing and technological aspects of desalination, but contained only one sentence indicating that brine disposal could be a major problem and that the ecological implications should be considered. More recently, the report on the past, present and future of desalination in Arab Gulf countries (General Secretariat Gulf Cooperation Council [GSGCC], 2014) describes the capacity per desalination technology in each country with projects listed up to 2011 and under study. However, the report does not mention energy sources used to power desalination plants, or environmental or sustainability aspects related to the future development of desalination capacity in that region. Ozair *et al.* (2017) acknowledge that there was an assumption that proper engineering would deal with negative impacts and that research was not addressing the environmental impacts of desalination. It was not until 2016 and 2017, that the World Bank, environmental authorities from Gulf Cooperation Council (GCC) countries and other stakeholders collaborated on the Gulf Environmental Partnership and Action Program (GEPAP), an initiative to address degradation of the environment in the Arab Gulf region (Arab News, 2017; Beatona, 2016; Ozair *et al.*, 2017). Nowadays, it is recognized that desalination involves trade-offs among the cost of building and maintaining desalinations plants, the benefits of providing fresh water for domestic, agricultural or industrial use, and the different environmental, economic and social impacts of the desalination plants (Figure 7.1).

The major environmental concerns for desalination plants have been related to the combustion of fossil fuels to produce the large amounts of energy required for the operating of the plants, and the discharge of concentrate (brine) as a by-product of the desalination process inland or in coastal areas (e.g., Al-Mutaz, 1991).

During the 1900s, the technology for desalination evolved and there are different alternatives with varying levels of energy consumption. Reverse osmosis (RO) is among the technologies that require less energy per m^3 of produced water. Manju and Sagar (2017) report that although in theory the energy required to desalinate a cubic metre of water (m^3) (3,405,000 parts per million (ppm)), is 0.86 kWh, the actual consumption is much higher. They report that desalination by multi-stage flash (MSF) distillation can use from 13.5 to 25.5 kWh/m^3, while RO requires from 2 to 5.5 kWh/m^3. Yet, the use of fossil fuels to power desalination plants continues to be a major environmental concern. In the Arabian Gulf, 69% of the installed capacity in 2008 corresponded to MSF evaporation and 23% to reverse osmosis (RO) (GSGCC, 2014). The International Energy Agency ([IEA], 2019) report that in the Middle East, where most of the capacity is installed, the production of water from desalination of seawater in 2016 still relied on the use

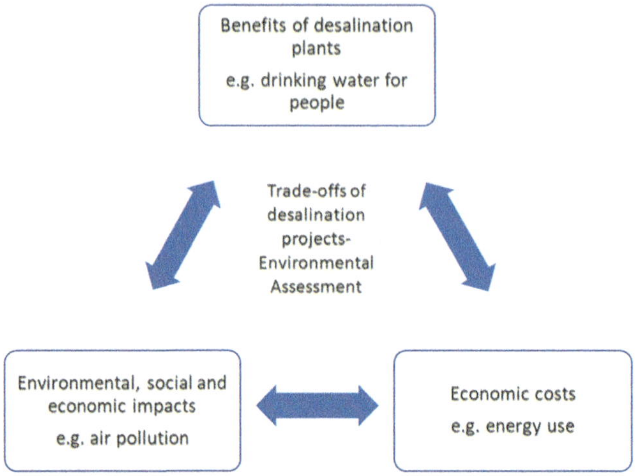

Figure 7.1 Trade-offs of desalination projects are determined through environmental and social assessment. (Source: authors' creation).

fossil fuels (e.g., natural gas, oil and others accounted for 99.1%). This contrasted with 0.7% produced with energy from renewable resources and 0.2% produced from nuclear energy. IEA (2019) estimates for 2040 consider that fossil fuels could still produce between 50 and 82%, renewables between 16.4 and 44.6% and nuclear between 3.5 and 8.7% depending on the scenario. Al-Shayji and Aleisa (2018) mention that in Kuwait, the main technology used is MSF desalination and attribute the reliance on fossil fuels to the government subsidies for electricity and conventional oil and gas products. MIDES has a purpose to reduce the cost and environmental impacts derived from the amount of energy needed to operate desalination plants. Although Global Water Intelligence (2020) maintains a detailed database of desalination plants under construction, presumed or online, it does not contain information about the source of energy. Therefore, a global study of environmental impacts of the emissions related to the existing and projected desalination capacity may still be needed.

The reduction in energy demand for desalination and the possibility of using renewable energy instead of fossil fuels or nuclear energy may help reduce some of the environmental impacts from emissions or the production of fossil fuels and nuclear energy. However, the discharge of brine is another major concern. The salinity of seawater varies from 32 to 37 g/L and in cold water could be less than 30 g/L (National Snow and Ice Data Center [NS&IDC], 2020). The salinity of brine may vary from 50 to 75 g/L (Lenntech, 2020) or approximately twice the salinity of seawater (Kenigsberg *et al.*, 2020). It is estimated that for the current 95 million m^3/day of desalinated water produced, around 142 million m^3/day of brine are discharged (Jones *et al.*, 2019). Brine has different effects on the marine environment and there are uncertainties about the short- to long-term effects on the aquatic and marine environment. Plants that use thermal technologies may release brine with a temperature higher than the temperature of seawater. The change in temperature and the discharge itself could modify water circulation and increase salinity especially in shallow locations (Ibrahim *et al.*, 2019) and could result in a decrease in abundance of diversity of foraminifera communities in benthic habitats (Kenigsberg *et al.*, 2020). Temperature changes are not a major concern for RO plants, but still the release of brine is. Cambridge *et al.* (2017, 2019) indicate that salinity of 54 g/kg could affect the metabolism, and consequently, the growth and survival of Sea grass (*Posidonia australis*), a key species in the Mediterranean Sea. Ozair *et al.* (2017) studied the effects of brine discharge from the Marafiq-Yanbu-I Desalination Plants in the Kingdom of Saudi Arabia and concluded, among other points, that there were slight differences in the productivity at the zone of the discharge compared with adjacent areas.

One of the limitations of desalination technology has been the high cost or energy required for the process. Caldera *et al.* (2018) report that the energy use for treatment of surface fresh water is around 0.6 kWh/m^3, while efficient RO plants require around 3 kWh/m^3 and thermal desalination plants from 40 to 80 kWh/m^3. The technological barriers are being addressed with the development of new technology. For example, MIDES aims to decrease the use of electricity for RO to 0.5 kWh/m^3 to reduce energy costs and the associated environmental and social impacts of energy generation. However, the environmental and social impacts are now getting more attention regarding the acceptance of projects, at least in the Western world. Irlbeck and Voutchkov (2013) comment that for the USA, the main barrier for many projects is no longer, the cost of energy for desalination, but the uncertainties regarding the impacts from desalination and how they are considered during the environmental approval process.

7.2 IMPACT ASSESSMENT FOR DESALINATION PROJECTS

Emissions resulting from the energy used and the impacts of brine release are two major concerns for the desalination process. However, the construction and operation of desalination plants have other social, environmental and economic impacts and benefits. The trade-off among those impacts and benefits is the foundation to determine the sustainability of desalination projects. Cooley *et al.* (2006) indicate that ignoring those trade-offs may jeopardize the financial feasibility of those projects or may result in their construction in areas where they cause significant environmental and social impacts. To prevent that, the impact assessment process, as defined by the International Association for Impact Assessment (IAIA), is an instrument to determine the positive and negative effects, intended or unintended, that projects, polices, plans, programmes and, in general, human interventions could have on the environment, (IAIA 2009). The impact assessment process engages the public, regulators and project proponents with the information needed to decide if a project is justifiable based on the expected benefits and impact, which become the economic, social, cultural and environmental costs or externalities.

The environment is defined in different ways in each country's legal instruments. In its broader sense, the environment is considered as composed of different parts and the interactions among all (modified from IAIA, 2009):

- Physical elements, for example land, water, soil, atmosphere, geology, and hydrology;
- Biological elements, for example aquatic and terrestrial fauna, flora and other organisms, including micro-organisms, ecosystems, and landscape;
- Socio-economic elements, including societal groups, cultural aspects related to each societal group, institutions, human rights, economy and livelihoods.

Similarly, laws and regulations define what is understood by 'impact' or 'effect', which often are used as synonyms. For example, the Impact Assessment Act of Canada (Government of Canada, 2020) defines effects as changes on the environment, health, social or economic conditions and the positive or negative results of those changes.

The practice of impact assessment has grown from its initial focus on the environment, which gave it the name of environmental impact assessment (EIA), to include socio-economic aspects, which is why in some cases it is called environmental and social impact assessment (ESIA) (see, for example, World Business Council for Sustainable Development [WBCSD], 2016), and this is the term that will be used in this chapter.

7.2.1 Existing guidance for the ESIA of desalination projects

Worldwide, the ESIA process follows specific steps, even when the specific requirements in the regulations may vary among countries or organizations that issue guidance. For example, the International Institute for Sustainable Development (IISD, 2020) considers 7 steps (Table 7.1) while UNEP (2007) considers 10 steps.

For this chapter we use a modified version of the 10 steps that the IAIA (1999) outlines. We add a pre-EISA step, project design, during which a multidisciplinary team can consider different options to reduce the overall environmental and social footprint of the project. For example, the decision of which desalination technology is better for a particular location should consider aspects such as the use of energy by different technologies (Figure 7.2), membrane or thermal processes, and the estimated emissions of greenhouse gases in addition to the economic costs. When thinking where to place the future desalination plant,

Table 7.1 Variation in steps of the ESIA process.

IISD (2016)	UNEP (2007)	IAIA (1999, Modified)
		(pre-ESIA) Project Design
Screening	Screening	Screening
Scoping	Scoping	Scoping
	Identification and description of policy and administrative aspects	
	Investigation and description of the proposed desalination project	
	Investigation and evaluation of environmental baseline	
		Definition of Alternatives
Impact assessment and mitigation	Investigation and evaluation of potential impacts of the project	Impact Analysis
Impact management	Mitigation of negative effects	Impact Management and Mitigation
	Summary and conclusions	
	Establishment of an environmental management plan	
		Determination of significance
Reporting		Reporting
Revision	Review of the EIA and decision-making process	Revision and Decision
Monitoring		Follow up

different locations can be compared to avoid impacts on sensitive areas, in addition to considering access to the site.

ESIA looks at the impacts considering the life of the project, typically divided into three phases: construction, operation and decommission or end of life; however, it can begin during the exploration and design (WBSCD, 2016). Citizens or residents can be involved in different parts of the process, as shown in Figure 7.2.

Each step of the process answers a specific question (Figure 7.2). In this chapter, the term 'effects' is used as a neutral word that refers to the positive and negative

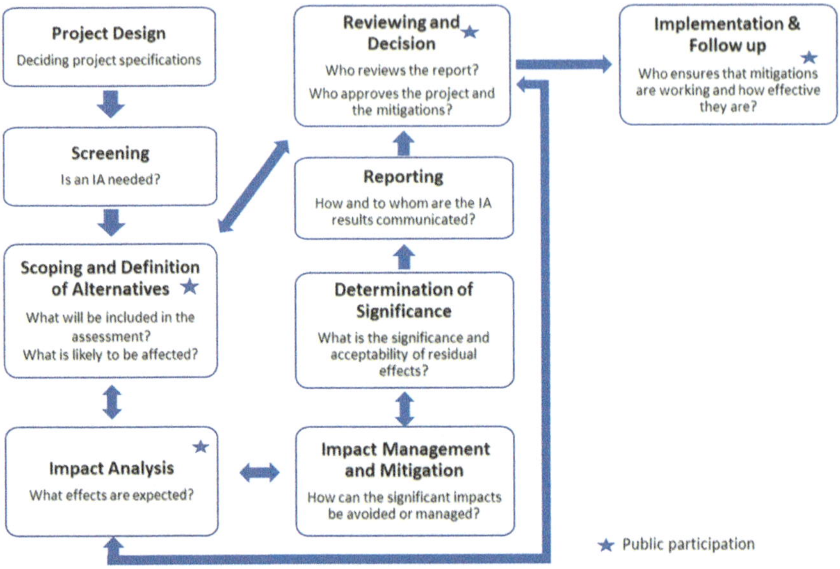

Figure 7.2 Steps in the process of environmental and social impact assessment. (Source: authors' creation).

changes that projects have on the environment. The term 'impacts', refers to the negative effects, and the term 'benefits' to the positive effects. During the scoping step, a key activity is to prepare the terms of reference that specify the scope or main impacts that should be covered in the assessment and other requirements that the assessment should comply with (WBCSD, 2016) and are specific to each project. In some countries, for example Canada, the environmental authority publishes the terms of reference to inform the public about the project and receive, from those interested, feedback on other aspects or concerns that should be added. In other countries the authority gives the terms of reference to the proponent to indicate what the final ESIA report should cover. During the impact analysis, the experts conducting the assessment need to consult the groups of people that can be impacted or benefited by the project, We merge the steps of reviewing and decision-making because in many countries, these steps are strongly linked, especially if, as in Canada, citizens are involved in the revision of the ESIA reports and/or a review panel is designated to review the ESIA and advise the Minister of the Environment (Government of Canada, 2020).

For the MIDES project, during this phase in the development of the technology, it is important to know what the expected impacts of the technology can be, in addition to its benefits.

Therefore, considering that the MIDES technology (MIDES hereafter) could be an integral part of a desalination project, using microbial desalination cells (MDCs)

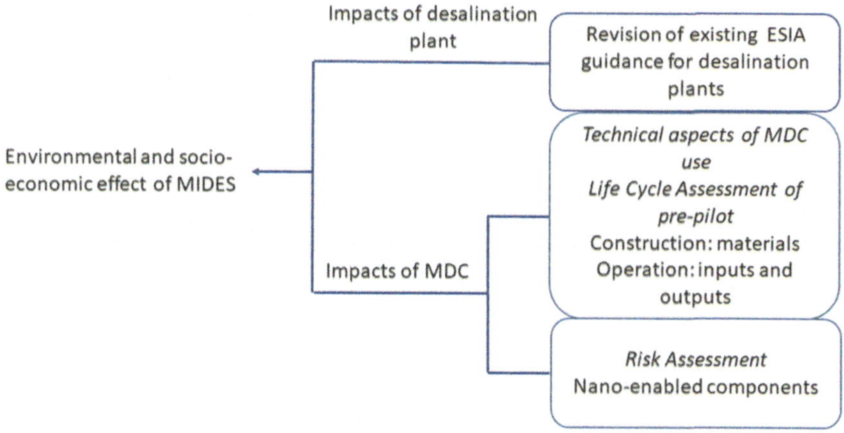

Figure 7.3 Assessment approach to determine the environmental and socio-economic effects of the MIDES technology. (Source: authors' creation).

as pre-treatment to reduce the salinity of seawater – or brackish water – before it goes into the RO unit, it is important to know what the environmental effects of that project would be.

An integrated approach was followed to identify the effects of MIDES (Figure 7.3). The first task was to determine what aspects were considered when analysing the effects of desalination projects. For that, a literature search was conducted to locate existing guidance for the ESIA of desalination projects and compile a list of impacts usually considered for this type of assessment (section 7.2.1.1). The second task was to identify the impacts associated with the MDC (section 7.3). Two different types of assessment were used to do that. Life cycle assessment (LCA) was used to identify and quantify combining three approaches. First, impacts associated with the production and operation of the microbial desalination cell were identified, based on literature on MDCs and microbial fuel cells. Second, because of the use of membranes and electrodes that contain nanomaterials, a risk assessment was conducted to determine the potential effects that would result from the release of nanoparticles. Third, LCA was used to identify and quantify the impacts associated with the production and operation of the MDC. Finally, the results of these tasks were integrated to create a comprehensive list of impacts or aspects to consider when conducting the ESIA for desalination plants that integrate MIDES (section 7.4).

7.2.1.1 Identification of impacts from existing ESIA guidance for desalination projects

The review of literature to identify guidelines for conducting an EIA for desalination projects started with a Google search using as key words, and in a different order, the terms EIA, Environmental Impact Assessment, Desalination, and Guide. Around 600 results of the search were reviewed to find 11 documents created by government agencies, international organizations or other groups (Table 7.2). The earliest guideline was in a form or an article by Hoepner (1999) published in the

Table 7.2 Main documents reviewed for impact assessment of desalination projects.

Author	Year	Guideline Name	Abbreviation
Hoepner, T.	1999	A procedure for environmental impact assessments (EIA) for seawater desalination plants	Hoep.
Ministry of Environment, Mauritius	2005	EIA Guidelines for Desalination Plants Mauritius	G. Mu
Swartz *et al.*	2006	A desalination guide for South African municipal engineers	S-ZA
World Health Organization	2007	Desalination for safe water supply	WHO
UNEP	2007	UNEP: Desalination Resource and Guidance Manual for Environmental Impact Assessments	UNEP, RGM
International Atomic Energy Agency	2010	Environmental Impact Assessment of Nuclear Desalination	IAEA
Monterey Bay National Marine Sanctuary (MBNMS)	2010	Guidelines for Desalination Plants in the Monterey Bay National Marine Sanctuary	MBNMS
PRODES	2010	PRODES: Guidelines for the regulation of desalination. Deliverable 6.2	PRODES
Water Research Foundation	2010	Guidelines for Implementing Seawater and Brackish Water Desalination Facilities	WRF
California Water Board	2015	Desalination Facility Intakes, Brine Discharges, and the Incorporation of other non-substantive changes	CWB
UNEP	2017	Updated guidelines on the management of desalination activities: UNEP(DEPI)/MED IG.23/23	UNEP. 23

journal *Desalination*. The earliest guideline created by an environmental authority was a word document by the Ministry of the Environment of Mauritius (2005). Four of the 11 documents were published in 2010. UNEP published a guideline in 2008 and an update in 2017 (see UNEP 2007, 2017).

During the scoping phase of an EISA, one of the main activities is to specify the baseline. This is the condition of the biotic, abiotic and socio-economic components of the environment. The revision of the guidelines produced 33 aspects of concern for 11 environmental components. These are organized in three types (Figure 7.4):

- Abiotic: aspects related to the changes in the quality of air, water, soil, and noise and vibrations.
- Biotic: aspects related to marine plants such as phytoplankton, algae, sea grasses and coastal vegetation communities such as marshes. Also, aspects related to the habitat, physiology, behaviour, reproduction or survival of marine invertebrates, fish, and marine mammals.
- Socio-economic: aspects related to economic activities that can be affected by the constructions or operation of the desalination plant, effects on cultural features or values, traffic and transportation on land or water, changes in the availability or demand for resources, induced growth of population or economic activities, and human health, both of workers and people living in the area of influence of the desalination plant.

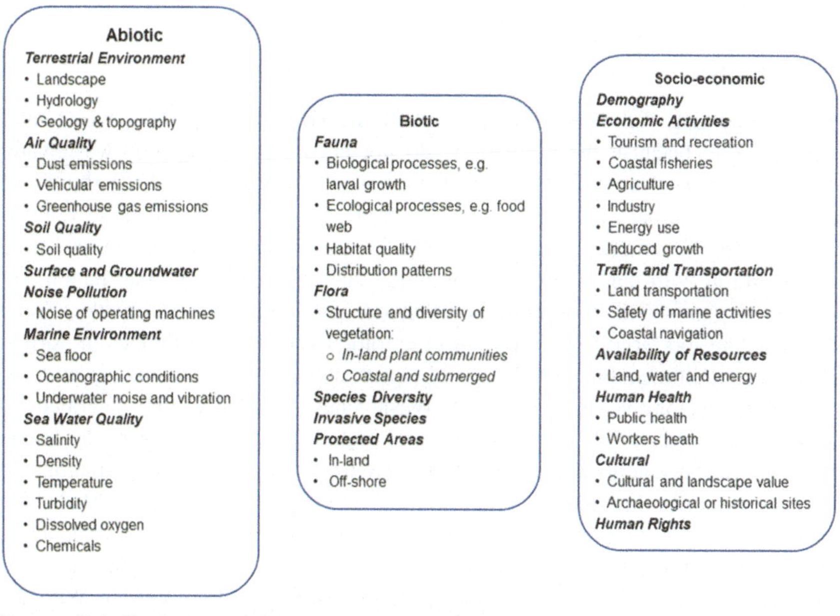

Figure 7.4 Environmental components and aspects for impact assessment of desalination projects. (Source: authors' creation).

Each one of these aspects has one or more impacts that should be analysed. An initial list of 133 impacts was compiled. To simplify, the list of impacts is integrated in the final list discussed in section 7.4. For example, during the construction of the plan, the quality of soil can be affected negatively in different ways. Impacts on soil can result from contamination with fuels or lubricants used by vehicles, but also from improper disposal of construction waste. During operation, soil contamination could happen if there is a leakage of chemicals used in the cleaning of the membranes.

7.3 IMPACTS ASSOCIATED WITH THE MICROBIAL DESALINATION CELL

The idea of using microbial fuel cells to reduce energy use in desalination was proposed by Cao *et al.* (2009). They suggested that the microbial fuel cells that were being developed to remove organic matter from wastewater could produce electricity to desalinate water. Research on MDCs has focused on scaling up the technology (e.g., Al-Mamun *et al.*, 2018; Guang *et al.*, 2020). Therefore, the identification of aspects related to the inclusion of the MDC as pre-treatment for reverse osmosis started with a review of literature to identify environmental and social concerns associated with the four key components of the MDC (Annex 1):

- Wastewater Intake System
- Ion Exchange Membrane (IEM)
- Cathode and Anode
- Catholyte

During the testing of the MDC-RO pilot plant, the MDC used synthetic wastewater as the anolyte because the regulations do not allow the use of real wastewater (Hernandez, pers. Comm.). The use of wastewater brings risks which should be analysed in an impact assessment. For example, the proximity of wastewater to the saline and desalinated water in the MDC raises concerns about the potential contamination with pathogens or compounds that could affect human health, for example organic and inorganic compounds, including heavy metals. Some molecules are too large to pass through the ionic exchange membranes in normal operating conditions, but could possibly pass and contaminate the desalinated water if the membranes are damaged. Wastewater, especially that generated by the food industry, is rich in organic matter, including volatile fatty acids. When these volatile fatty acids accumulate in the environment, they feed methanogenic microbes, contributing to the generation of methane (Park *et al.*, 2018) and eutrophication. Organic matter, especially volatile fatty acids are removed from wastewater in the MDC. This will help reduce the risk of eutrophication in receiving water bodies; however, other compounds may still be present. With the use of synthetic wastewater for testing MIDES it was not possible to compare the composition of the water discharged from the MDC with the composition of the wastewater used as the anolyte.

To determine if additional treatment will be required to reduce the risk of contamination by the treated wastewater, it is recommended that the pilot plants are tested with real wastewater and that the substances in the treated wastewater are monitored. This will help determine the presence of substances that could contribute to water pollution before it is disposed of, for example into surface waterways or the sea, or before it is used for other purposes such as irrigation or replenishment of groundwater.

The biofilm in the membranes of the MDC is formed by *Geobacter sulfurreducens*. Besides its use in microbial fuel cells, this bacterium has been proposed as an agent to recover silver and other metals from wastewater (Law *et al.*, 2008), but it also has been reported to cause corrosion of steel (Mehanna *et al.*, 2009). Because this is a new technology, the risk of bacteria escaping from the MDC if the membranes are damaged is unknown. If the bacteria escape, potentially this could, in the long term, contribute to corrosion of steel pipelines if the bacteria form colonies in them. The cleaning of the membranes and the use of pulse flow and reverse flow may help prevent this risk.

Ceramic membranes are used as pre-treatment of seawater or brackish water. The ion exchange membranes are used in the anode and cathode of the MDC. Both types of membranes are already used in desalination and the impacts commonly associated with them derive from the use of chemicals for membrane cleaning and the disposal of membranes and other materials at the end of their useful life (see section 7.2 and Annex 1).

MIDES tested two different types of cathode at laboratory scale: an air diffusion cathode made of carbon nanofibres and iron nanoparticles, which used oxygen reduction, and a liquid cathode that used ferro-ferricyanide redox (Ramirez-Moreno *et al.*, 2019). The pilot plant used the liquid cathode. We considered both cathodes to identify potential impacts. For the air-cathode, the main concern is the toxicity that iron or manganese oxide (alternative) nanoparticles can have on the environment and on human health. For example, the inhalation of nanoparticles of manganese oxide is reported to cause irritation, neurotoxic effects and hypoventilation (Zaitseva *et al.*, 2015).

For the liquid cathode, ferrocyanide ions can react in the environment in the presence of light and be lethal to fish even at 1 ppm (Burdick & Lipschuetz, 1950). They can also react in acid medium and at high temperature to produce hydrogen cyanide, which is highly toxic. Therefore, proper ventilation and personal protective equipment should be used when handling ferrocyanide solutions (Michaud, 2017).

The risks for the release of nanoparticles to the environment are not specified in guidelines for the ESIA of desalination plants. Therefore, one of the tasks of MIDES was to conduct a nano-safety risk assessment. The results are described in the next section.

7.3.1 Risk assessment of nano-enabled components

Engineered nanomaterials (ENMs) are natural or manufactured materials that contain at least 50% of unbound, aggregate or agglomerate particles that, in one or more of their external dimensions, measure from 1 to 100 nanometres (nm); this includes, for example fullerenes, graphene flakes and single-wall carbon nanotubes (European Commission [EC], 2011). Nanoparticles are usually composed of existing substances but when the molecules are extremely small in size (nano form), a certain mass in nano-form would display different properties (behaviour) than the same mass of the same substance composed of macroscopic-size molecules. For example, they can have a surface area or chemical reactivity several orders of magnitude larger than the macroscopic form (Kabir *et al.*, 2018). For example, Singh *et al.* (2013) compared the toxicological effects of micro and nanoparticles of manganese oxide and found that the nanoparticles had toxic effects at smaller concentrations per body weight than the microparticles (300 vs 1000 mg/kg, respectively). This different behaviour is one of the causes of concern for nanoparticles. For example, exposure of rats to nanoparticles of iron oxide (Fe_2O_3) resulted in damage to lung and liver tissue (Sadeghi *et al.*, 2015). Iron nanoparticles were associated with decreases in the enzyme super oxide dismutase in the brain and liver of fish (Shaw & Hardy, 2011).

In water treatment technology, ENMs with diverse chemistry (i.e., carboneous materials, metals and metal oxides, magnetic core composites) have been used as absorbents or coagulants, and to improve membrane filtration (permeability, mechanical properties), to create conductive devices and sensors, to obtain photocatalytic property, to improve the Fenton and photo-Fenton reaction, as well as for pollutant remediation (Adeleye *et al.*, 2016). ENMs are used to significantly improve the performance and reduce cost and for applicability of water treatment technology (Adeleye *et al.*, 2016; Bai *et al.*, 2017). However, potential release of ENMs can occur, by direct injection of ENMs into the environment (i.e., in-situ groundwater treatment) (Bossa *et al.*, 2017a) or indirectly during product cleaning (Bossa *et al.*, 2017b).

Leitat Technology Centre was responsible for the risk assessment for the MDC based on the engineered nanomaterials (ENMs) incorporated in it. The MDC process developed in MIDES uses the principle of electrodialysis. Two differently polarized electrodes extract most of the sodium and chloride ions from the seawater (treated water). These ions migrate through two ion exchange membranes into two separate chambers and cannot go back to the original chamber. Largely salt-free water remains in the middle chamber, which is then completely desalinated through reverse osmosis, requiring considerably less energy compared to state-of-the-art technologies. In an initial stage, MIDES partners worked on the development and selection of the materials to be further tested in a lab-scale prototype. The development of the MDC considered an air-cathode and a liquid-cathode. From these, only the materials of the air-cathode incorporate ENMs. Based on the product performance, solely the cathode containing iron nanoparticles and one cathode incorporating manganese oxide, developed by SGL, were selected for scaling up the technology.

Exposure to ENMs during the use of the air-cathode was assessed using a lab-scale experiment to simulate a classic use of the air-cathode for desalination. The potential hazard of released iron and manganese was estimated by literature review.

7.3.1.1 Assessment of ENMs exposure

The MIDES project used an ENMs risk assessment framework (Hristozov *et al.*, 2016) to estimate the probability that adverse human health or environmental effects could emerge from exposure to ENMs. Identifying whether the substance is dangerous involves determining both its potential toxicity and also the possibility and degree of exposure (Wiesner *et al.*, 2009). The exposure assessment describes how a chemical is used by workers and consumers and to what degree the ENM is released into the environment to estimate the routes of exposure (i.e., inhalation, ingestion or dermal) and the involved environmental compartments (e.g., water, sediment, soil) and engineering compartment (e.g., Warheit, 2018) in the factory (i.e., far and near-field), in the wastewater treatment plant or in landfill. Exposure estimation to assess the quantity and form of ENMs reaching humans can involve the use of models, in-situ measurement using the analytics technique and/or lab experiment simulation of the activities occurring during the ENMs life cycle. Ideally, every new product introduced to the market that incorporates ENMs (i.e., a nano-enabled product (NEP)), should undergo a risk assessment to ensure it is safe for the consumer and the environment. The hazard assessment involves the analysis of available data on (eco) toxicological effects to establish dose-response relationships. The characterization of risk combines hazard and exposure.

7.3.1.2 Release during use

The release of ENMs from the air-cathode was tested at lab scale. A sample of the catholyte was analysed to detect the potential transfer of ENMs from the air-cathode through the flash-spun high-density polyethylene membrane into the recirculating catholyte (800 mM $K_3Fe(CN_6)$ and 100 mM phosphate-buffered saline [PBS] buffer) (Figure 7.5).

The cation exchange membrane (CEM) located between the catholyte and the saline chamber (treated water) was also analysed to detect any potential retention of ENMs that were inside the pores or stuck to the surface of the cation exchange membrane (Figure 7.5). The details of the lab-scale experiment for the air-cathode are detailed in Table 7.3.

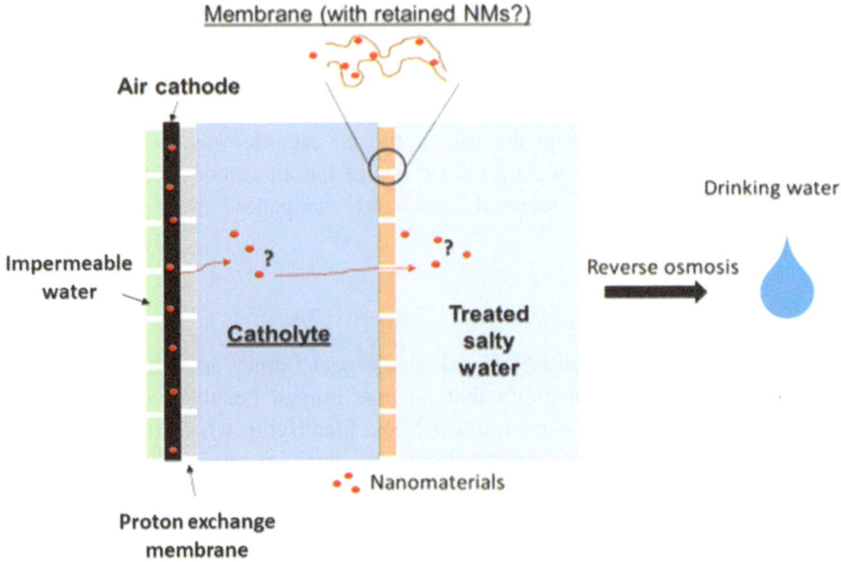

Figure 7.5 Setup of the air-cathode for desalination. (Source: authors' creation).

7.3.1.3 Analysis of the catholyte solution

The strategy to analyse the release of ENMs in the catholyte follows a decision tree to determine the analyses required (Figure 7.6). Starting with the total release (particulate and ionic fraction), the next step is to differentiate soluble from the release of particles. This continues with the determination of the form and speciation of particles released and concludes with the assessment of their reactivity. Each previous step must be validated to pursue the strategy. Chemical analysis was performed by inductive plasma mass spectrometry/atomic emission spectroscopy (ICP-MS). Prior to the analysis, the sample of the catholyte suspension was filtrated at 10 kDa, using a stirred ultrafiltration (UF) cell of 76 mm diameter to concentrate the particulate fraction. The concentrated factor obtained was about 100. Samples of the liquid (initial and concentrated catholyte) and the membrane were digested with acid in an analytical microwave digestion system (microwave acid digestion system [MARS], CEM Corporation, 160 W). The amount of manganese and iron transferred from the air-cathode to the catholyte and the membrane was determined by ICP-MS (Agilent 7500, Agilent Technologies).

7.3.1.4 Degradation of the air-cathode during use and hazard assessment

The air-cathode was analysed by SEM before and after use to characterize its integrity after being used. Characterization of the ENMs present at the surface of

Table 7.3 Experimental conditions for Fe and Mn air-cathode at each sampling time.

Parameters	Desalination Cycle	Passive Desalination	Catholyte MDC 3 mM Fe Cathode (4 L)	Catholyte MDC MnO2 (MC2) (2.5 L)
Contact time (days)	7	20	1:1 mix catholyte (21 d: 35 d)	7
Cation exchange membrane (CEM)	Type 10	Type 10	Type 10	Type 10
Circulation flow rate (L/h)	5, 7	7	5, 7	5, 7
Chambers volume relation (Anolyte: Saline: Catholyte)	(5:1:5)	(5:1:5)	(5:1:5)	(5:1:5)
Current collectors	Stainless steel	Stainless steel	Stainless steel	Stainless steel
Cathode	CNF's + [Fe] = 3 mM	CNF's + [Fe] = 3 mM	CNF's + [Fe] = 3 mM	GDL (SGL) + $[MnO_2]$ = 23.66 g
Anode	Inoculated carbon felt (KFD 2.5)	Non inoculated carbon felt (KFD 2.5)	Inoculated carbon felt (KFD 2.5)	Inoculated carbon felt (KFD 2.5)
[NaCl] initial (g/L)	Seawater 1:4	Seawater 1:4	Seawater 1:4	10
Applied load (Ω)	2.50	–	2.50	2.50
Status	Long term operation Delivered	Long term operation Delivered	Long term operation Delivered	Long term operation Delivered

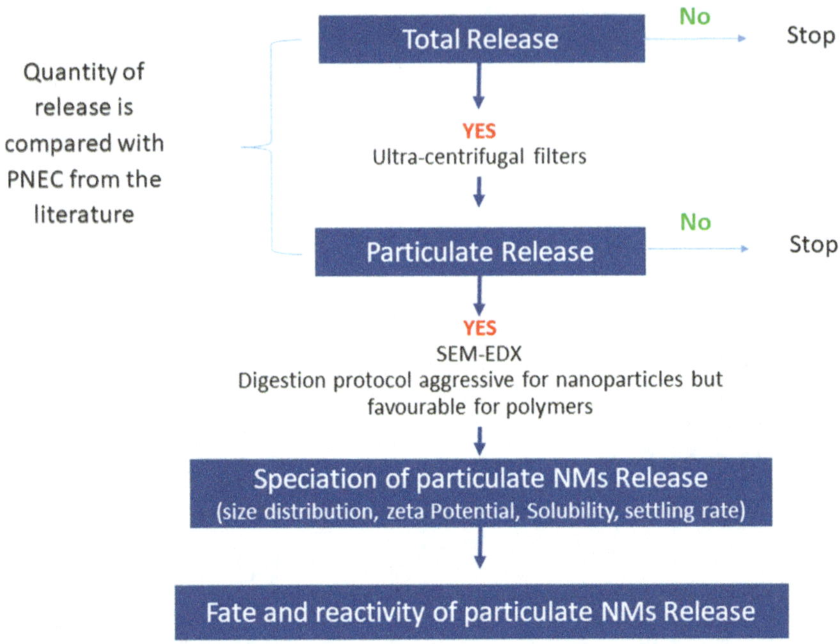

Figure 7.6 Decision tree of sampling analysis strategy. (SEM: scanning electron microscopy; EDX: energy dispersive X-ray spectroscopy) (Source: authors' creation).

the air-cathode matrix before and after was performed by visual inspection. EDX analysis was used to determine the content of ENMs before and after use of the air-cathode.

The concentration and form of the released nanomaterials was determined to assess the risk associated with them. The predicted concentration was compared with the predicted no-effect concentration extracted from the literature. An effort was made to extract data from toxicological and ecotoxicological internal projects and public resources such as papers or reports.

7.3.1.5 Analysis of ion exchange membranes

Leitat fabricated carbon nanofibres doped with iron nanoparticles for use in the air-cathode of the MDC. During the search for literature, no articles were found specifically on the release of ENMs from electrospun carbon nanofibres, which shows the need to study this aspect further. The SEM investigation showed no change in the morphology of the nanofibres, indicating that the integrity of the ENMs is preserved. Studies by Alegre *et al.* (2017) and Wang *et al.* (2017) also found electrospun carbon nanofibers to be stable. However, these studies only assessed membrane integrity after a short period of use. In MIDES, the use time

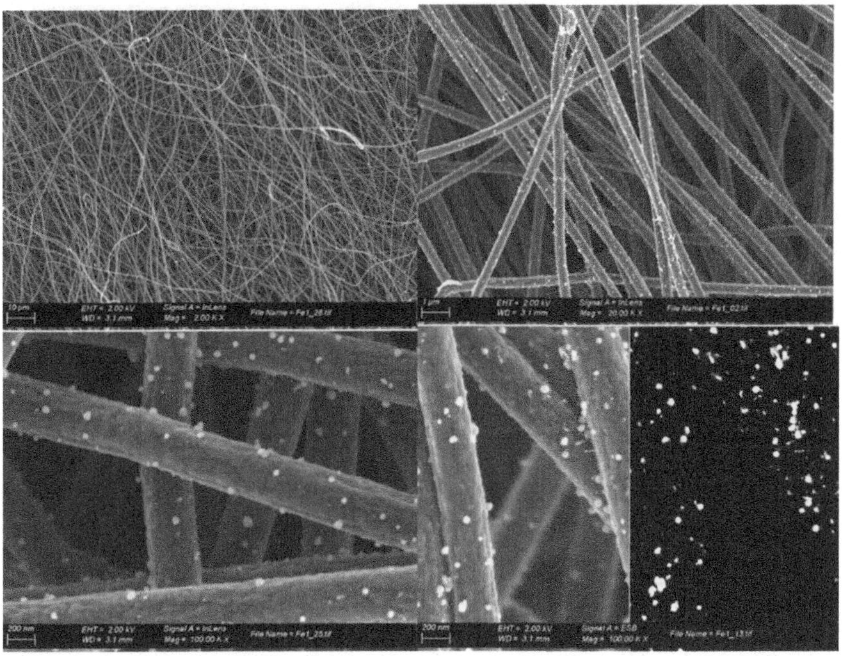

Figure 7.7 SEM images of new iron air-cathode. (Source: authors' creation).

of the aged air-cathode was 7 days. SEM investigation of new and aged iron ENMs is presented in Figure 7.7 (new, Fe cathode); Figure 7.8 (aged, Fe cathode); Figure 7.9 (new manganese cathode) and Figure 7.10 (aged, manganese cathode).

The iron air-cathode is composed of intermixed carbon nanofibres (200 nm in diameter). The iron ENMs are visible as bright spots on the SEM images. Backscattered-electron (BSE) images reveal the presence of well dispersed Fe ENMs and larger aggregate. The BSE images of the new Fe air-cathode confirm the presence of Fe ENMs. On the contrary, BSE images of the used Fe air-cathode do not reveal the presence of iron ENMs. It should be noted that in the SEM image only the surface of the cathode is visible. There is no information on the iron ENMs present deeper in the carbon nanofibres. The iron content quantified by SEM-EDX confirms the visual inspection: that there is a decrease in the iron content from 11.2 ± 0.6 when new to 5.9 ± 1.6 w.% when aged. However, no decrease of iron content by ICP-MS was detected, meaning that iron release is limited to the surface of the air-cathode. In a broader perspective, the carbon nanofibres appear to be compacted and their surface when aged shows a decrease in the roughness of the surface, appearing smoother than when they were new. This could result from the compaction of the cathode during use.

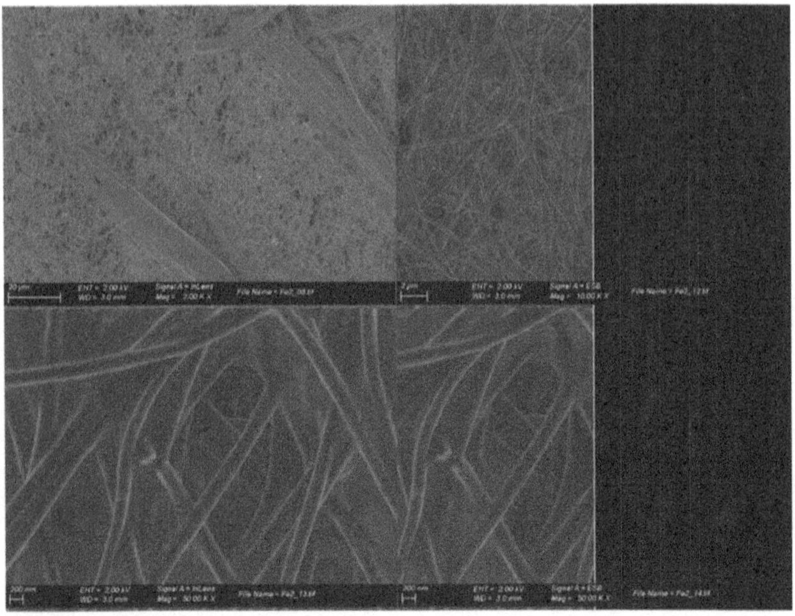

Figure 7.8 SEM images of aged iron air-cathode. (Source: authors' creation).

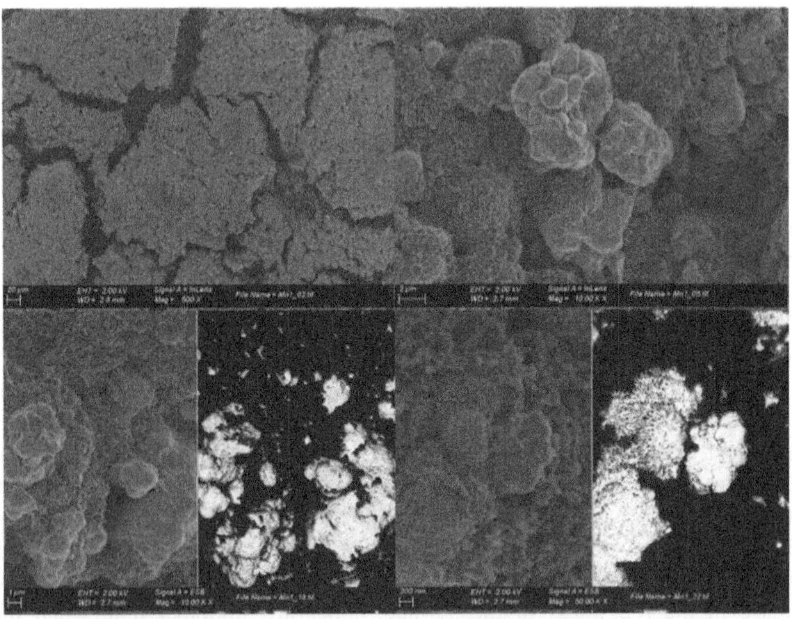

Figure 7.9 SEM images of new manganese air-cathode. (Source: authors' creation).

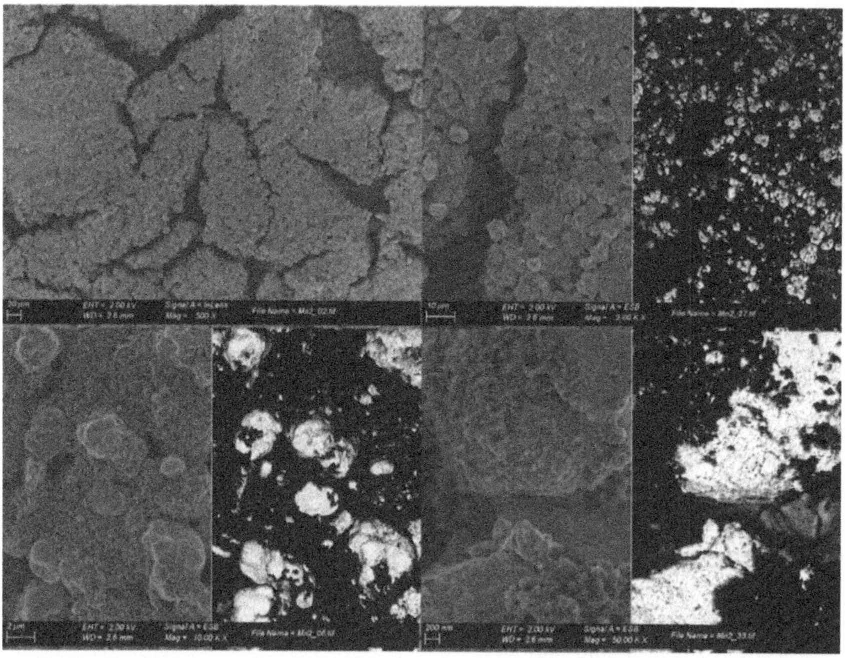

Figure 7.10 SEM images of used manganese air-cathode. (Source: authors' creation).

The manganese cathode is composed of a support layer and a catalyst layer of carbon black and MnO_2. This manganese cathode was made by the MIDES partner SGL. Differing from the Fe cathode, the MnO_2 particles are mostly present aggregated and are found to be mixed with porous carbon black materials. On low magnification it was observed that the catalyst layer was not fully uniform; the presence of ravines allows observation of the carbon support layer. The total concentration of manganese quantified by ICP-MS was 4.3 w.% and by SEM-EDX analysis it was 24.1 ± 3.7 w.%. This difference is because SEM-EDX is a surface analysis of about 5 microns depth and the support layer without MnO_2 was excluded from this analysis. In BSE images, the size of the MnO_2 particles varied from a few micrometres to the nanometric range. The content of manganese quantified by SEM-EDX decreased significantly, from 24.1 ± 3.7 w.% for the new cathode, to 15.7 ± 6.2 w.% after aging (i.e., use). However, similarly to the Fe carbon nanofibres from the cathode, no decrease was observed by ICP-MS. This means that the decrease in manganese is limited to the surface of the cathode (i.e., 5 μm depth). In contrast, to the Fe carbon nanofibres from the cathode, the visual inspection by SEM did not show any difference between the new and aged manganese air-cathode.

7.3.1.6 Release of ENMs from the air-cathode

The release of ion and manganese ENMs was assessed during the use of the air-cathode. The air-cathode is separated on the outside by a proton exchange membrane and by an impermeable membrane on the inner side (Figure 7.5). The impermeable membrane lets the air pass through but not the water molecules, and any release through it is unlikely to occur. On the other side, the proton exchange membrane allows the passing of cations and protons to balance the charge and allows a certain moisture to be maintained. ENMs detached from the air-cathode could potentially diffuse through the proton exchange membrane into the catholyte and then through the membrane separating the catholyte and the treated salty water. Just by analysing the setup of the air-cathode in the MDC, it appears unlikely that the ENMs could diffuse through the two membranes into the treated water. In this way, the sampling was first focused on the catholyte and the membrane separating the catholyte and the desalinated water (cation exchange membrane) to analyse any ENMs released from the air-cathode into the catholyte and trapped in the membrane's pore network or at the membrane's surface, respectively. Sampling was carried out after conducting the set of experiments shown in Table 7.3. It should be noted that four samplings were performed with the materials of the iron air-cathode and only one for the manganese air-cathode.

During all sampling periods for the catholyte, the concentration of iron and manganese detected by ICP-MS was below the limit of quantification, that is 1.2 ppb for manganese and 30.4 ppb for iron (calculated using the blank value + 5 times the blank standard deviation). To lower the limit of quantification, we concentrated the particulate fraction by ultrafiltration to remove the ionic species and concentrate the particulate fraction. The concentration factor was between 50 and 100 times. In concentrated solutions, the iron measured by ICP-MS was always below the limit of quantification. For the sampling with the manganese air-cathode, the quantity of manganese measured in the concentrated catholyte was insignificant (0.45 ppb after applying the concentrating factor). The potential release of iron from the air-cathode appears to be limited but the amount of manganese particles released was within detectable limits.

The total concentration of iron in the new and used (after desalination experiment) cation exchange membrane did not show any significant difference. No release of ENMs from the surface of the membrane or from inside the membrane pore network was detected. For all desalination cycles, the decision tree (Figure 7.6) was stopped after analysis of the released particles. Iron exposure during the use of the iron air-cathode appears limited, the release of iron ENMs was visually detected on the surface of the membrane; however, this amount was too low to be detected by ICP-MS. For the configuration of the manganese air-cathode, the release of manganese was detected in the concentrated solution, revealing some potential exposure.

7.3.1.7 ENMs hazard assessment

The released concentration of nanoparticles was estimated to be up to 0.45 ppb for manganese and 30.4 ppb for iron (limit of ICP-MS quantification). We compared these data with the literature for environmental and human toxicity. For human toxicity we focus on oral exposure as potential ingestion through treated water was expected.

For the assessment of iron toxicity, the toxicity values for FeO and FeOx are used because the speciation of iron is unknown. The ecotoxicity of iron ENMs, particularly nanoscale zero-valent iron (nZVI) materials (FeO), has been evaluated mostly with a focus on aquatic and terrestrial ecotoxicity, concentrations from 0.5 mg/L to above 2.5 g/L have been reported to be toxic (El-Temsah *et al.*, 2016; Keller *et al.*, 2012; Saccà *et al.*, 2014). Zhang *et al.* (2016) found that after 72 hours of exposure to particles (from 0.86 to 2.27 μM), equivalent to concentrations of 0.05 to 0.13 mf of iron per litre, there was an effect on green algae. They observed accumulation of the nanoparticles in the gastrointestinal track of Daphnia, with low toxic effects, with exposure to uncoated nanoparticles of iron oxide (374 to 1,181 μM), equivalent to 21 to 66 mg of iron per litre (mg Fe/L),

For human toxicity, legal threshold values for human consumption are set to 200 μg/L for iron and 50 μg/L for manganese (Environmental Protection Agency, 2014). It can be seen that the threshold for iron is much higher than the limit reported for ecotoxicity of nanoparticles of iron oxides. However, Harford *et al.* (2014) proposed a trigger value to act on manganese toxicity at 45 μg per litre, while Peters *et al.* (2010) suggested a predicted no effect concentration (PNEC) of 62 to 123 μg/litre. For manganese nanoparticles, Singh *et al.* (2013) found a toxic effect on rats at concentrations of 300 mg/kg of body weight. The effects included damage to the DNA of cells in the bone marrow, inhibition of aminotransferases in liver, kidneys, and also detected inhibition of acetylcholinesterase in the brain.

7.3.2 Life cycle assessment of microbial desalination cells

Life cycle assessment (LCA) is a standardized technique to provide understanding of the impacts associated with the life cycle of manufactured products, from the materials used in the production process, to their use and final disposal or end of life (ISO, 2006). The results of the LCA can be used to improve a product or technology to reduce the environmental impacts associated with its inputs (material, energy) and outputs (resources, emissions, etc.) and to inform decision makers about its environmental performance (ISO, 2006). LCA has been used to analyse the potential impacts that can derive from different aspects of the desalination process, for example emissions of greenhouse gases, or materials and chemicals used (Cherif & Belhadj, 2018), or to compare alternatives for the disposal of RO membranes at their end of life (Lawler *et al.* 2015). For the MDC,

Faze (2015) and Zhang *et al.* (2018), used LCA to determine the impacts of lab scale, single desalination cells (one chamber, with anode and cathode) and both concluded that the MDC was a suitable technology but it required better efficiency and improvement in the design and materials to be viable. Faze (2015) indicated that the impacts included emissions impacting climate change and toxicity for marine organisms, and contribution to global warming, while Zhang *et al.* (2018) indicated the manufacturing and operation of the MDC had the larger impacts.

The LCA for the MIDES project followed the phases indicated in ISO (2006):

- Definition of the goal, scope, and system boundaries;
- Life cycle inventory analysis;
- Impact assessment; and
- Interpretation.

7.3.2.1 Goal and scope

For the LCA the system boundaries were limited to the MDC, using a fully functional pre-pilot (laboratory scale) model built and tested by IMDEA. An attributional LCA from cradle-to-gate was carried out, following the guidelines of ISO 14 044 (ISO 2006) and the International Reference Life Cycle Data System Handbook (European Commission-Joint Research Centre-Institute for Environment and sustainability, 2010). The goal for the LCA was to estimate the potential impacts resulting from the pre-pilot MDC stack construction (material inputs) and use (energy use). The pre-pilot MDC system was composed of 12 unit cells (unit cell: anode/saline compartment/cathode) and it was operated at the laboratories of IMDEA, one of the project partners (Figure 7.11). The functional unit was the production of 1 m^3 of desalinated water from seawater.

As the anolyte, the MDC used pre-treated wastewater from a local brewery to reduce the chemical oxygen demand (COD) from 1,933 to 240 mg/L. Synthetic seawater (sodium chloride [NaCl]) was used for the experiments. As the catholyte, the MDC used potassium hexacyanoferrate ($K_3[Fe(CN)_6]$) and iron (III) chloride ($FeCl_3$).

7.3.2.2 Life cycle inventory and inventory analysis

OpenLCA 1.74 (Green Delta) was used to run the LCA. The ReCiPe Midpoint Hierarchist (H) V1.13 method was used because the review of literature showed that it is commonly used for desalination studies (e.g., Faze, 2015; Opher& Friedler, 2016). ReCiPe calculates 18 impact categories (Table 7.4).

The inventory included materials used for the construction of the MDC. The laboratory tests done by IMDEA, complemented with data from the database Ecoinvent 3.5, were used to specify the amounts of inputs (e.g., electricity, catholyte, and anolyte) and outputs (desalinated water, treated water, dissipated energy).

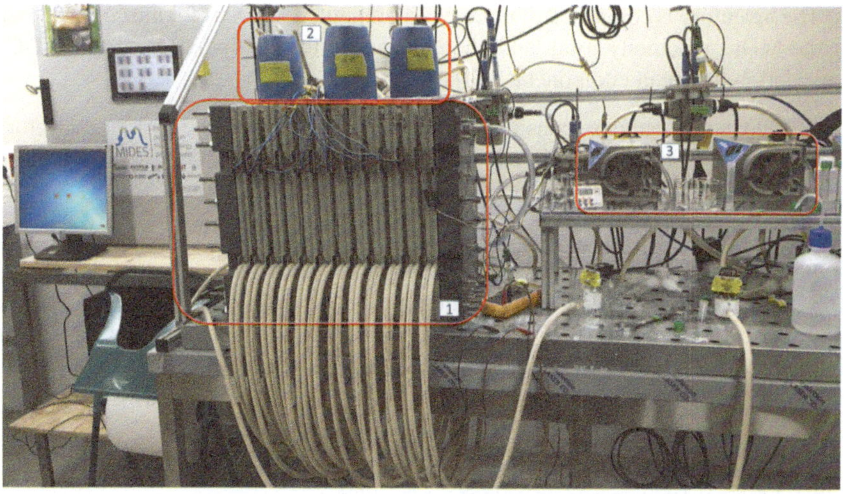

Figure 7.11 Pre-pilot MDC stack. 1 = MDC stack, 2 = conductivity probes, 3 = recirculation pumps. (Source: IMDEA).

Table 7.4 Impact categories estimated by the ReCiPe method.

Abbr.	Impact Category	Reference Unit
HT	Human Toxicity	kg 1,4-DCB-Eq
MRD	Metal Depletion	kg Fe-Eq
CC	Climate Change	kg CO_2-Eq
FD	Fossil Depletion	kg oil-Eq
MET	Marine Ecotoxicity	kg 1,4-DCB-Eq
FET	Freshwater Ecotoxicity	kg 1,4-DCB-Eq
IR	Ionizing Radiation	kg U_{235}-Eq
WD	Water Depletion	m^3 water-Eq
TA	Terrestrial Acidification	kg SO_2-Eq
ULO	Urban Land Occupation	m^2a
ALO	Agricultural Land Occupation	m^2a
POF	Photochemical Oxidant Formation	kg NMVOC-Eq
PMF	Particulate Matter Formation	kg PM10-Eq
FE	Freshwater Eutrophication	kg P-Eq
ME	Marine Eutrophication	kg N-Eq
TET	Terrestrial Ecotoxicity	kg 1,4-DCB-Eq
OD	Ozone Depletion	kg CFC-11-Eq
NLT	Natural Land Transformation	m^2

The inventory is composed of three types of inputs: materials for the construction of the MDC, electricity used, and chemicals used for the operation of the MDC, considering the main components of the MDC:

- Anode and cathode, considering for both the chamber, the current collector, the electrode and the silicone joint;
- Saline chamber;
- Anion and cation exchange membranes with silicone joints;
- Other materials (miscellaneous): tanks, pumps, pipes, valves, screw and bolts.

The complete inventory and assumptions, together with the detailed LCA, are reported by Jamtsho (2019). Table 7.5 presents the main assumptions from literature, and inputs and outputs for a scenario of complete desalination, in which the MDC is used to reduce salinity from 35 to 0.5 gNaCl/L (seawater to

Table 7.5 Assumptions and data used for the LCA.

Parameter	Value	Reference
Functional unit	1 m^3 of desalinated water	–
Surface of membrane (polyamide)	40.9 m^2	–
Iron (III) Chloride	95.75 kg/m^3	–
Life span of the MDC stack	20 years	–
Life span of ion exchange membranes	5 years	Hancock *et al.* (2012), Tarnacki *et al.* (2011)
Ratio desalinated water: anolyte:catholyte	1:20:20	
Operating pressure	83 bar	–
Life span of plant	20 years	Garfí *et al.* (2016), Muñoz and Fernández-Alba (2008)
Chemical oxygen demand (COD) (anolyte input)	1993.00 mg/L	–
Chemical oxygen demand (COD) (anolyte output)	240.00 mg/L	–
Hours of operation per year	8000	–
No. of pumps in the MDC	3	–
Power (recirculation pump)	21.34 W/pump	–
Dissipated energy	0.69 kWh/m^3	–
Energy recovery rate of pressure exchanger	40%	–

drinking water). Because of limitations in the database, ferro-ferricyanide was substituted with iron (III) chloride.

7.3.2.3 LCA interpretation

ReCiPe uses 18 categories to calculate the impacts associated with the amounts of inputs and outputs. The results (Table 7.6) show that the construction of the MDC, including the replacement of components during the 20 years considered, has large contributions in three impact categories (Table 7.6): human toxicity, with 3262.59 kg of 1,4-dichlorobenzene equivalents (DCB-Eq), metal depletion, with 607.77 kg of iron equivalents (Fe-Eq), and climate change, with 594.48 kg of carbon dioxide equivalents (CO_2-Eq). There were low contributions to depletion of fossil fuels (199.25 kg of oil equivalents), marine ecotoxicity (56.62 kg 1,4-DCB-Eq), fresh water ecotoxicity (53.67 kg 1,4-DCB-Eq), and ionizing radiation (38.26 kg of $Uranium_{235}$ equivalents [U_{235}-eq]). The contributions to the other impact categories were negligible.

The results of the LCA (Table 7.7) show that for the 20 years considered, the impacts associated with the construction of the MDC are negligible when compared to the impacts associated with its operation. Among these, it is possible to observe that the use of the catholyte (ferro-ferricyanide, substituted by $FeCl_3$), is the main contributor. The largest benefit results from the treatment of wastewater (WWT) and there are minor benefits related to the reduction in energy required for RO and the energy dissipated from the MDC that could potentially be used in other parts of the RO process.

7.4 INTEGRATED ASSESSMENT FOR MIDES-DESALINATION PROJECTS

MIDES uses microbial desalination cells (MDCs) as a pre-treatment step for reverse osmosis (RO), simultaneously desalinating a saline stream and treating wastewater. This combination offers the possibility to reduce the energy needed to desalinate

Table 7.6 Main contributions to impact categories.

Abbr.	Impact Category	Reference Unit	Result
HT	Human Toxicity	kg 1,4-DCB-Eq	3262.59
MRD	Metal Depletion	kg Fe-Eq	607.77
CC	Climate Change	kg CO_2-Eq	594.48
FD	Fossil Depletion	kg oil-Eq	199.25
MET	Marine Ecotoxicity	kg 1,4-DCB-Eq	56.62
FET	Freshwater Ecotoxicity	kg 1,4-DCB-Eq	53.67
IR	Ionizing Radiation	kg U_{235}-Eq	38.26

Table 7.7 Results of the life cycle impact assessment for the partial and complete desalination scenarios.

Impact[1]	Units	MDCc[2]	Energy I[3]	$FeCl_3$[4]	WWT[5]	Energy II[6]	Electr.[7]	Total
Climate Change	kg CO_2-Eq	<0.01	7.74	78.38	−21.06	−1.52	−0.30	63.24
Fossil Depletion	kg oil-Eq	<0.01	2.23	22.60	−4.78	−0.44	−0.09	19.54
Human Toxicity	kg 1,4-DCB-Eq	0.01	5.42	80.95	−7.96	−1.07	−0.21	77.15
Marine Ecotoxicity	kg 1,4-DCB-Eq	<0.01	0.17	2.03	−0.42	−0.03	−0.01	1.74
Metal Depletion	kg Fe-Eq	<0.01	0.08	11.20	−0.79	−0.02	−0.00	10.45
Water Depletion	m^3 water-Eq	0.00	0.09	1.54	−2.23	−0.02	−0.00	−0.68

Notes: 1 = impact category, 2 = MDC construction, 3 = electricity for pumps (recirculation), 4 = iron (III) chloride, 5 = avoided wastewater impacts, 6 = avoided electricity for seawater RO, 7 = electricity dissipated (avoided).

seawater. Because it is a new technology, the adoption of MIDES requires an understanding of its benefits and impacts, before it is scaled-up for commercial use, to inform citizens and decision-makers. The approach used by MIDES partners integrated three methods. Fist, a review of existing guidance for the environmental and social assessment (ESIA) of desalination projects was conducted to compile elements that are considered in terms of reference and best practice for ESIA. Then, the impacts that could derive from the MDC were identified by a literature review on impacts associated with the components of the MDC such as membranes, anode and cathode. This was followed by the use of a risk assessment to evaluate environmental and social risks from the release of nanoparticles from the components of the MDC that include nanomaterials (cathode and ion exchange membranes). Finally, life cycle assessment (LCA) was used to estimate the impacts and benefits associated with the construction and operation of the MDC considering its inputs – such as feed water, anolyte, materials, chemicals and energy – and outputs – such as desalinated water, treated wastewater and dissipated energy. The information obtained from the three methods was used to compile a new list of elements that could help draft terms of reference for the ESIA of desalination projects that integrate MIDES.

7.4.1 Revision of existing guidance for environmental and social impact assessment

The main environmental impacts specific to desalination relate to five characteristics of the process:

- The large amount of energy needed for removing the salt from seawater or brackish water (either for thermal or membrane methods);
- The use of chemicals for pre-treatment of water (flocculation and coagulation processes) and cleaning the membranes;
- The construction and operation of the intake of water from the source to the desalination plant;
- The construction and operation of the outlet to discharge of brine into the environment; and
- The disposal of used membranes.

Other impacts related to the construction of the plant, such as use of land, traffic of vehicles during construction, noise, and excavation, or other aspects related to operations (e.g., generation of solid and human waste, use of vehicles, or energy usage for facilities) are similar to the impacts considered for the assessment of other types of projects.

The revision of 13 guidance documents for the environmental and social impact assessment (ESIA) of desalination projects showed that no single guideline covered all 23 environmental components identified, including aspects to which MIDES could contribute positively or negatively (Figure 7.12). For example, the guide

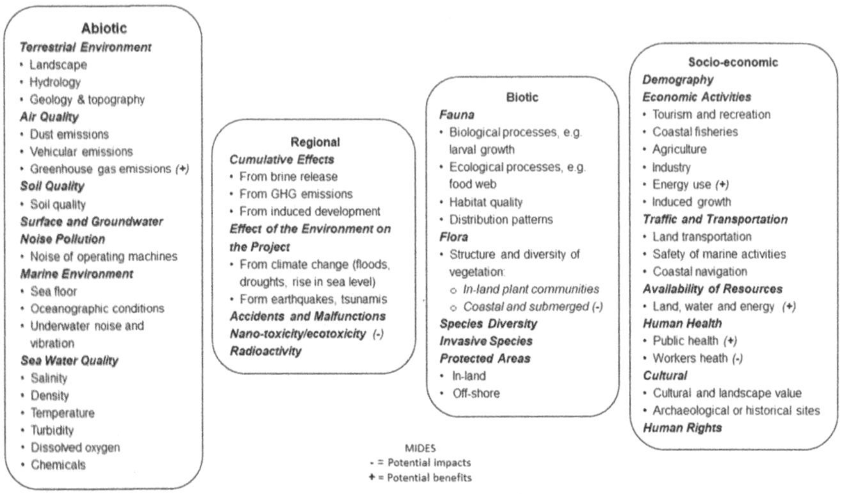

Figure 7.12 Environmental and social impacts identified for desalination projects that integrate MIDES. (Source: authors' creation).

created by the UNEP (2007, 2017) covered environmental and social aspects but not all that were identified in our compilation. The guide created by the World Health Organization (WHO, 2007) covered some environmental aspects but focused on human health impacts. However, the guides did not specifically mention human rights. Access to water is a human right and assessing how a desalination project can improve access to water could increase the social acceptance of projects that may be controversial. There may be cases in which human rights should be considered for other reasons. If land needs to be expropriated for the project, it is important to assess if this has been done following principles that guide impact assessment practice, for example do not harm, right to consultation and right to proper compensation.

In the compilation there were regional aspects that were not covered properly and are considered gaps in guidance: the effects that the environment could have on the desalination project, and regional effects to which the project could contribute (cumulative effects).

The Arab Gulf is one of the world regions in which nuclear power is used for desalination (GSGCC, 2014). The use of nuclear power for desalination is expected to increase (World Atomic Association, 2020); however, concerns about radioactivity were only mentioned in the guide by the International Atomic Energy Agency (2010). This agency specifies the needed to ensure that the coupling of desalination and nuclear plants follows strict safety standards to avoid radioactive contamination of desalinated water. Tritium is a hydrogen radioactive isotope created in fusion reactors. It is considered highly soluble and it is present in water and is not removed by RO membranes.

7.4.1.1 Effects of the environment on desalination projects

Best practices for EIA/ESIA include the estimation of risks associated with accidents and malfunctions, and the effect that the environment could have on a proposed project (e.g., Environmental Protection Agency, 2017). In the reviewed guidelines, little attention was given to these risks, such as the effects that the environment could have on desalination plants. Some risks may have attached a certain probability of occurrence, for example the time of return of a storm of a given magnitude. But for others this is not easy to determine, for example for the occurrence of earthquakes of a magnitude that can damage infrastructure.

With the expected increase in severity of climate phenomena such as hurricanes or floods, the increase in sea level, and earthquakes, it is important to assess how these events could affect desalination infrastructure because of the selection of a location or other factors. This is important to consider because it could result in social impacts on the population that would depend on the water produced by the plant, affecting the supply of drinking water and/or water available for other uses, for example food production or urban irrigation. Another effect could be an impact (negative) on human health and the economy of the region served by the plant when drinking water is not available even for a short time. The expected changes in climate patterns vary across world regions and the intensity of impacts may increase when combined with the impacts of land degradation by human activities (IPCC, 2019). To protect desalination infrastructure, climate change predictions should be assessed in conjunction with other expected trends such as increases in land development or water supply and demand.

7.4.1.2 Cumulative effects of desalination projects

Another aspect that is not covered properly is the contribution to cumulative effects that desalination projects can have. While the guidelines integrated air pollution, only five guidelines considered climate change none of which were from the cumulative perspective. In the desalination literature, cumulative effects are not commonly mentioned. With most of desalination capacity still relying on fossil fuels to generate energy for the process (IEA, 2019), a major concern is the contribution that greenhouse gas (GHG) emissions resulting from the use of fossil fuels have on air pollution and climate change. For example, desalination plants in Kuwait are estimated to contribute to GHG with 2.71×10^{10} kg CO_2 eq. and 1.15×10^{8} kg SO_2 eq. annually, in addition to large amounts of other air pollutants such as CFC, 1,4DB, C_2H_4, and PO_4 (Al-Shayji & Aleisa, 2018).

Brine discharge can result in cumulative effects that are not yet explored in the literature, although studies are starting to show the sensitivity of various species to increased salinity. In the Mediterranean Sea, sea grass *Podsonia australis*, a key species in the ecosystem, has reduced growth when exposed to changes in salinity (Cambridge *et al.* 2019), and laboratory experiments indicate that anti-scalants and brine can have synergistic cumulative effects on reef-building corals due to osmotic shock, combined with eutrophication (Petersen *et al.*, 2018). Given the expected increase in desalination projects (e.g., GSGCC, 2014), concerns about the cumulative effects of desalination projects on aquatic organisms and communities in gulfs and other marine and coastal environments may exacerbate.

Regional impacts associated with desalination, such as the potential cumulative effect due to the discharge of brine on seas and gulfs, and the effects of the environment and climate change on the desalination infrastructure, are not captured properly at the level of project assessment. They should be addressed at a regional level, for example through strategic environmental assessment. However, this impact assessment tool has not been used commonly for desalination.

7.4.2 Environmental and social impacts and benefits of MIDES

Most of the impacts of MIDES are the same as those of other RO desalination plants, for example, the environmental impacts include potential water pollution from the use of chemicals for cleaning membranes, if the residual water from the RO process is not treated before being discarded in waterbodies. Only nine potential impacts and three benefits were directly related to the MDC. The following sections describe the main environmental and social impacts and benefits identified for MIDES.

7.4.2.1 Environmental Impacts

The environmental impacts exclusive to the MDC are related to the production and use of the catholyte. The results, using the pre-pilot unit, showed high contributions to climate change and ecotoxicity. However, the first results of the operation of the pilot plant in Denia showed a significant reduction in the ratio of produced water to catholyte. During the operation of the pilot MDC, the feed water to catholyte ratio decreased from 1:20 to 1:5. This essentially reduces the contribution to these two impact categories, getting closer to a point where the benefits could surpass the impacts and make the MDC a suitable pre-treatment for reverse osmosis desalination.

7.4.2.1.1 Air pollution and contribution to climate change

Contribution to climate change (e.g., global warming potential or CO_2 eq.) was considered in the LCA with focus on the main technological components or processes included in the design of a desalination project (e.g., Hancock *et al.* 2012). Considering a period of 20 years, the LCA indicated that the impacts associated with the construction of the MDC were negligible. The main impact categories to which the operation contributed were climate change, human toxicity and fossil depletion. The emissions associated with the catholyte were the main contributors to climate change. The main contribution to all the impact categories is the catholyte, also contributing to climate change (594.48 kg of CO_2 eq.) and depletion of metals (extraction of natural resources), with 607.77 kg of iron. This negative impact results from the use of the catholyte which currently is not reconstituted after use. When comparing the use of an air diffusion cathode and a ferro-ferricyanide redox cathode for the MDC, Ramirez-Moreno *et al.* (2019) concluded that the air-cathode is a viable alternative for desalination of brackish water, but the liquid cathode (ferro-ferricyanide) is, for now, the viable option for desalinating seawater. This highlights the need to keep improving the efficiency of the MDC, especially by reducing the ratio of produced water to anolyte and catholyte (w:c:a). The pilot plant already shows a significant improvement in the efficiency. For the pre-pilot used for the LCA, the w:c:a ratio for seawater was 1:20:20 and for brackish water it was 1:15:15. The first results of the operation of the pilot plant in Denia, Spain, show a ratio of 1:5:5. This is closer to the 10-fold reduction that the sensitivity analysis of the LCA identified as the operation ratio with which the emissions avoided by the use of the MDC would surpass the emissions generated (and contribution to climate change) by the production of the catholyte. With the 10-fold reduction the contribution to toxicity would be minimal and the use of the MDC would result, overall, in more benefits than impacts. The best alternative to minimize the impacts of the MDC is to regenerate and reuse the catholyte in the process. Although this requires more research and the regeneration would require other inputs (e.g., energy), this improvement would magnify the benefits of the MDC.

7.4.2.1.2 Nanoparticles and ecotoxicity

Environmental and social impact assessment (ESIA) reports should include an analysis of risks of accidents and malfunctions or disasters (Environmental Protection Agency, 2017). However, these aspects were not commonly considered in the guidelines reviewed and none mentioned nanomaterials. This may be in part because the impacts of nanomaterials were not well known a decade ago (Wiesner *et al.*, 2009). For MIDES, accidents and disasters need to be integrated, adding the need to consider the risks associated with the potential release of nanoparticles in the environment. The lab experiments run to determine the release of nanoparticles from the ion exchange membranes and the catholyte indicated that there is limited release of nanoparticles during the use of the air-cathode, and that the release of manganese nanoparticles is below the threshold for water consumption and the risk of nanotoxicity is negligible. The potential for contamination of soil or water with nanoparticles because of disposing of membranes on landfills should be minimized by looking for other methods to dispose of membranes or preferably reusing or recycling them.

Although the lab experiments for the MDC indicate that there is limited release of iron nanoparticles from the surface of the membranes, during the use of the air-cathode a release of manganese nanoparticles was detected, but at a concentration below the threshold for water consumption by a factor of 100. In this condition, the risk associated with the release of nanoparticles is low. However, the results obtained at the lab-scale may differ from the real conditions when the larger (pilot) MDC or a full commercial version are used. Given the expected increase in desalination plants (e.g., Manju & Sagar, 2017) and the increase of ion exchange membranes in desalination (Ran *et al.*, 2017), the concerns about ecotoxicity due to nanoparticles may not be exclusive to the MIDES technology.

Potassium hexacyanoferrate used in the pre-pilot as the catholyte is toxic to aquatic organisms (Santa Cruz Biotechnology, 2010). Storage and handling according to instructions in datasheets should be followed to avoid spills that could contaminate water bodies.

7.4.2.1.3 Release of treated wastewater

The impacts that the existing guides need to incorporate to cover MIDES technology are related to the use and disposal of the wastewater (anolyte). The pilot plant was tested with synthetic wastewater because the use of real wastewater was not allowed (Hernandez, Pers. Comm.). To reduce concerns of potential contamination of drinking water by the treated wastewater, it would be convenient to test, at least at the level of pre-pilot, the MDC with real wastewater to determine the compounds in the released treated wastewater. This will help determine the presence of substances that could contribute to water pollution before it is discharged into surface waterways, the sea or used to replenish groundwater. The presence of nanoparticles in the wastewater treated in the MDC was not considered in the risk assessment, although it was estimated that the release of nanoparticles from the surface of the membranes and the cathodes was negligible. Nevertheless, it would be important to monitor the presence of nanoparticles in the residual water discarded.

7.4.2.2 Social impacts

The social impacts of MIDES identified with the LCA and risk assessment consisted mainly of toxicity resulting indirectly from the amount of catholyte used and potential exposure to nanoparticles or the liquid catholyte.

7.4.2.2.1 Human toxicity

Human toxicity is a social-health impact that could result when humans are exposed to nanoparticles released from nano-enabled materials. Although it is a concern for human health, the information on toxicity of nanoparticles has focused mostly on aquatic organisms or laboratory animals, for which toxic effects have been identified. Manganese oxide was not included in the LCA because it was not used in the pilot plant. However, if it were to be used, it could have associated impacts for human and environmental toxicity. There could be a concern, especially for accidental exposure of workers involved in the fabrication of the membranes (Zaitseva *et al.*, 2015). The release of iron or manganese nanoparticles to water or land could be an indirect effect depending on how the membranes are handled after their use in the MDC – or in the desalination plant – and depending on the disposal method. The recycling of membranes is the best option to reduce the impacts deriving from their construction and disposal, although this is still not a common practice (J. Senam, Pers. Comm.).

The toxicity associated with the liquid catholyte was estimated by substitution of potassium hexacyanoferrate with iron chloride. This was a limitation of the database and it is possible that this impact was overestimated. Potassium hexacyanoferrate is not considered a hazardous substance for humans. Its effects on human health may consist of discomfort or irritation of mucous membranes, eyes or skin if there is direct contact with the skin or eyes or it is inhaled (Santa Cruz Biotechnology, 2010). However, its thermal or chemical reaction in acid conditions may result in the production of hydrogen cyanide, which is lethal (Santa Cruz Biotechnology, 2010). These negative impacts are considered low and can be mitigated by following standard handling procedures: storage and use in acid-free environments and use of personal protective equipment should be required to handle this substance.

Another potential impact on human health, identified as one of the impact categories in the LCA, is related to the production of 1,4 Dichlorobenzene (DCB), which is considered a carcinogenic compound. In the LCA it may represent other substances with similar carcinogenic effects that are associated with inputs and outputs of the construction and operation of the MDC. This impact can be considered indirect and may also be overestimated in the LCA because of the substitution of potassium hexacyanoferrate with iron (III) chloride, which is corrosive. Manganese oxide was not included in the LCA because it was not used in the pilot plant. However, if it were to be used, it could have associated impacts for human and environmental toxicity. The former could be a concern especially for accidental exposure of workers involved in the fabrication of the membranes (Zaitseva *et al.*, 2015).

7.4.2.3 Environmental benefits

The main environmental benefits of MIDES derive from the reduction in energy usage for the whole desalination process and the reduction of COD in the wastewater used as anolyte. The environmental benefits of the MDC result from its bio-electro-chemical technology, in which biological wastewater treatment is integrated into the desalination of a saline stream using ion exchange membranes without external energy input. MDCs simultaneously treat wastewater and perform desalination using the energy contained in the wastewater. In fact, an

MDC can produce around 1.8 kWh of bioelectricity from the energy contained in 1 m^3 of wastewater. Compared to traditional RO, the MDC saves more than 3 kWh/m^3.

7.4.2.3.1 Reduction in energy demand for desalination and reduction in GHG emissions

While with MIDES there are emissions associated with the construction and operation of the MDC, and the pre-treatment of wastewater before its use in the MDC, the pilot plant in Denia shows improvement in the operation compared to the pre-pilot. Most of the emissions in the LCA are related to the catholyte, with minor contributions from the energy needed for the recirculation pumps. However, the benefits identified for the MDC include the energy savings, which may be larger if the dissipated energy can be stored or used in other parts of the system. The improvement in efficiency shown by the pilot plant in Denia, by reducing the ratio of feed water to catholyte used from 1:20 to 1:5, indicates that the energy usage for desalination can be reduced.

A feasibility study for a desalination plant in Australia (Melbourne Water, 2007) indicated that a 150 GL desalination plant would use annually around 90 MW of electricity, resulting in about 1 million tonnes of CO_2 emissions. The operation of the pilot MDC is still being optimized, so future data will show the actual contribution to the reduction in energy and emissions from desalination.

7.4.2.3.2 Treatment of wastewater and reduction in eutrophication potential

The wastewater used as anolyte has a high content of volatile fatty acids. The release of this wastewater in the environment would contribute to eutrophication of water bodies. The amount of anolyte is proportional to the amount of catholyte used. The LCA showed that the treatment of wastewater results in avoided impacts in the categories of climate change, depletion of fossil fuels, marine toxicity and depletion of water. This means that by using the organic matter in the wastewater, the MDC can have the benefit of a reduction in the use of fossil fuels to generate energy for wastewater treatment. This results in other benefits: the reduction in emissions related to the use of energy for treatment of wastewater and the reduction in eutrophication potential. The latter could result in improvements in the habitat quality for aquatic organisms.

7.4.2.4 Social benefits

The positive social impact of MIDES includes direct and indirect benefits.

7.4.2.4.1 Reducing operating cost of desalination

According to estimates from 2007, for a desalination plant that would produce 200 million m^3 of water annually, the energy used in the desalination process can represent more than 50% of the operating costs. Using renewable energy, the annual energy usage of a desalination plant may exceed 100 million Australian dollars (Melbourne Water, 2007). By reducing the amount of electricity needed for desalination by pre-treating seawater with the MDC, desalination can be more affordable in low-income countries with access to seawater or brackish water and for which desalination is still too costly. Besides the technological aspects, the installation of the pilot plant in Denia showed that water and health authorities were concerned about the health risks that could result from the use of wastewater in the pre-treatment of sea water. More work is still needed to demonstrate that the process is safe and to remove any legal barriers that could delay the integration of the MDC at a larger scale.

7.4.2.4.2 Increase in supply of drinking water and improvement in water and sanitation

Affordable desalination would improve access to drinking water in locations where seawater or brackish water is available. Improvements in access to drinking water and in hygienic and sanitary conditions can help water authorities fulfil the access to clean water and sanitation as a human right and as a goal for sustainable development, improving the quality of life of people in those regions.

7.4.2.4.3 Contribution to reuse of wastewater and use of aquatic species

The other direct benefit is to help reduce the costs and environmental impacts of treating wastewater that could be used as the anolyte in the MDC. By reducing the COD in wastewater, the MDC can help increase the amount of treated water available for other uses, for example irrigation or industry. This can help reduce the pressure on fresh water resources. Indirect benefits from the reduction in COD especially in countries with limited treatment of wastewater would be the reduced discharge of organic matter in the environment and eutrophication that affects water bodies. This can lead to improvements in the quality of surface water and in the habitat of aquatic organisms such as fish. These improvements, in turn, would benefit communities depending on fish and other aquatic resources in aspects such as nutrition and livelihood.

7.5 CONCLUSION

The use of microbial desalination cells as pre-treatment for reverse osmosis makes MIDES a technology able to help simultaneously reduce two problems in the water sector: the scarcity of drinking water and the discharge of wastewater that contaminates water bodies. Using the chemical energy in wastewater to reduce the salinity of seawater and brackish water, MIDES also helps reduce the environmental impacts that result from the generation of large amounts of energy required for desalination.

Most of the impacts of MIDES are the same as those that are associated with desalination plants, for example, the environmental impacts include potential water pollution from the use of chemicals for cleaning membranes, or changes in water quality and organisms in marine environments.

The environmental impacts exclusive to the MDC are related to the production and use of the catholyte. The results, using the pre-pilot unit, showed high contributions to climate change and ecotoxicity. However, the first results of the operation of the pilot plant in Denia already show a significant reduction in the ratio of produced water to catholyte. During the operation of the pilot MDC, the feed water to catholyte ratio decreased from 1:20 to 1:5. This largely reduces the contribution to these two impact categories, getting closer to a point where the benefits could surpass the impacts and make the MDC a suitable pre-treatment for reverse osmosis desalination.

The risk assessment for nanoparticles indicated that the release of nanoparticles from the ion exchange membranes and the catholyte was in concentrations below the threshold for water consumption; consequently, the risk of ecotoxicity on the environment was considered negligible. However, the use of ion exchange membranes is increasing in desalination indicating the need to consider nanotoxicity as a potential risk to analyse and monitor for desalination projects in general, not exclusively for MIDES projects. The presence of nanoparticles in the residual water from a full-size MDC should be monitored to validate the results obtained in the laboratory and ensure that, if there is a release of nanoparticles, it is within safe limits for human and ecosystem health. The catholyte is toxic in the aquatic environment but is not considered toxic to humans, although it can cause irritation if there is direct contact. These negative impacts can be mitigated by following standard handling procedures and checking the integrity of the membranes.

The environmental benefits of the MDC result from its bio-electro-chemical technology, in which biological wastewater treatment is integrated into the desalination of a saline stream using ion exchange membranes without external energy input. With this novel technology, two low-quality water streams (saline stream, wastewater) are transformed into two high-quality streams (desalinated water, treated wastewater) suitable for further uses.

The positive social impact of MIDES could be expressed by several direct and indirect benefits. One direct benefit is that, by reducing the amount of energy needed to desalinate water, MIDES can make RO affordable in countries with access to seawater or brackish water and for which desalination is still too costly. Another direct benefit is to help reduce the costs and environmental impacts of treating wastewater that could be used as the anolyte in the MDC. The increased supply of water could lead to indirect benefits such as improvements in access to drinking water and in hygiene and sanitary conditions. The reduction in organic compounds in wastewater could also contribute to reduction in eutrophication potential. This could benefit the habitat quality for aquatic organisms and fish, which could also benefit people depending on those ecosystems.

REFERENCES

Adeleye A. S., Conway J. R., Garner K., Huang Y., Su Y. and Keller A. (2016). Engineered nanomaterials for water treatment and remediation: costs, benefits, and applicability. *Chemical Engineering Journal*, **286**, 640–662.

Alegre C., Busacca C., Di Blasi O., Antonucci V., Aricò A. S., Di Blasi A. and Baglio V. (2017). A combination of CoO and Co nanoparticles supported on electrospun carbon nanofibers as highly stable air electrodes. *Journal of Power Sources*, **364**, 101–109.

Al-Mamun A., Ahmad W., Baawain M. S., Khadem M. and Dhar B. R. (2018). A review of microbial desalination cell technology: configurations, optimization and applications. *Journal of Cleaner Production*, **183**, 458–480.

Al-Mutaz I. S. (1991). Environmental impact of seawater desalination plants. *Environmental Monitoring and Assessment*, **16**, 75–84.

Al-Shayji K. and Aleisa E. (2018). Characterizing the fossil fuel impacts in water desalination plants in Kuwait: a Life Cycle Assessment approach. *Energy*, **58**, 681–692. https://doi.org/10.1016/j.energy.2018.06.077.

Apec Water. (2017). History of Reverse Osmosis Filtration. https://www.freedrinkingwater.com/reverse-osmosis/knowledge-base/history-of-reverse-osmosis-filtration.htm.

Arab News. (2017). Jeddah hosts workshop on Gulf environmental degradation. January 31. https://www.arabnews.com/node/1046941/saudi-arabia.

Bai L., Bosssa N., Qu F., Winglee J., Li G., Sun G. L., Lang H. and Wiesner M. R. (2017). Comparison of Hydrophilicity and Mechanical Properties of Nanocomposite Membranes with Cellulose Nanocrystals and Carbon Nanotubes. *Environmental Science & Technology*, **51**, 253–262.

Beatona. 2016. Kuwait organizes workshop on GEPAP. June 7. http://www.beatona.net/en/news/kuwait-organizes-workshop-gepap.

Bossa N., Carpenter A. W., Kumar N., De Lannoy C.-F. and Wiesner M. (2017a). Cellulose nanocrystal zero-valent iron nanocomposites for groundwater remediation. *Environmental science Nano*, **4**, 1294–1303

Bossa N., Chaurand P., Levard C., Borschneck B., Miche H., Vicente J., Geantet J., Aguerre-Chariol O., Michel M. and Rose J. (2017b). Environmental exposure to TiO_2 nanomaterials incorporated in building material. *Environmental Pollution*, **220**, 1160–1170.

Burdick G. E. and Lipschuetz M. (1950). Toxicity of Ferro- and Ferricyanide Solutions to Fish, and Determination of the Cause of Mortality. *Transactions of the American Fisheries Society*, **78**(1), 192–202.

Caldera U., Bogdanov D. and Breyer C. (2108). Desalination costs using renewable energy technologies. In: Renewable Energy Powered Desalination Handbook Application and Thermodynamics, V. Gnaneswar Gude (ed.), Elsevier, Ondford, UK, Cambridge MA, USA, pp. 287–329. ISBN: 978-0-12-815244-7. https://doi.org/10.1016/B978-0-12-815244-7.00008-8.

California Water Board. (2015). Amendment to the Water Quality Control Plan for Ocean Waters of California Addressing desalination facility intakes, brine discharges, and the incorporation of other non-substantive changes Division of Water Quality State, Water Resources Control Board, California Environmental Protection Agency. http://www.waterboards.ca.gov.

Cambridge M. L., Zavala-Perez A., Cawthray G. R., Mondon J. and Kendrick G. A. (2017). Effects of high salinity from desalination brine on growth, photosynthesis, water relations and osmolyte concentrations of seagrass *Posidonia australis*. *Marine Pollution Bulletin*, **115**, 252–260.

Cambridge M. L., Zavala-Perez A., Cawthray G. R., Stalton J., Mondon J. and Kendrick G. A. (2019). Effects of desalination brine and seawater with the same elevated salinity on growth, physiology and seedling development of the seagrass *Posidonia australis*. *Marine Pollution Bulletin*, **140**:462–471.

Cao X., Huang X., Liang P., Xiao K., Zhou Y., Zhang X. and Logan B. E. (2009). A new method for water desalination using microbial desalination cells. *Environmental Science & Technology*, **43**, 7148–7152.

Cherif H. and Belhadj J. (2018). Environmental life cycle analysis of water desalination processes. In: Sustainable Desalination Handbook. Plant Selection, Design and Implementation, V. Gnaneswar Gude (ed.), Elsevier, Butterworth-Heineman, Cambridge, MA, USA, pp. 527–559.

Cooley H., Gleick P. H. and Wolff G. (2006). Desalination, with a Grain of Salt A California Perspective. Pacific Institute for Studies in Development, Environment, and Security, Oakland, Calif. USA.

El-Temsah Y. S., Sevcu A., Bobcikova K., Cernik M. and Joner E. J. (2016). DDT degradation efficiency and ecotoxicological effects of two types of nano-sized zero-valent iron (nZVI) in water and soil. *Chemosphere*, **144**, 2221–2228.

Environmental Protection Agency, Ireland. (2017) Guidelines on the information to be contained in environmental impact assessment reports. Draft, August. https://www.epa.ie/pubs/advice/ea/EPA%20EIAR%20Guidelines.pdf

Environmental Protection Agency, Ireland. (2014). Drinking Water Parameters Microbiological, Chemical and Indicator Parameters in the 2014 Drinking Water Regulations 2014. An overview of parameters and their importance. https://www.epa.ie/pubs/advice/drinkingwater/2015_04_21_ParametersStandaloneDoc.pdf.

European Commission. (2011). Commission recommendation of 18 October 2011 on the definition of nanomaterial. (2011/696/EU).

European Commission - Joint Research Centre - Institute for Environment and Sustainability. (2010). International Reference Life Cycle Data System (ILCD) Handbook - General guide for Life Cycle Assessment - Detailed guidance. First edition March 2010. EUR 24708 EN. Luxembourg. Publications Office of the European Union.

Faze N. R. (2015). Life Cycle and Economic Analysis Comparing Microbial Desalination Cell and Reverse Osmosis Technologies. MSc, Ohio State University.

Garfí M., Cadena E., Sanchez-Ramos D. and Ferrer I. (2016). Life cycle assessment of drinking water: comparing conventional water treatment, reverse osmosis and mineral water in glass and plastic bottles. *Journal of Cleaner Production*, **137**, 997–1003.

General Secretariat, The Cooperation Council for the Arab States of the Gulf [GSGCC]. 2014. Desalination in the GCC. The History, the Present & the Future, 2nd edn. The Cooperation Council for the Arab States of the Gulf General Secretariat. Riyadh, Saudi Arabia. https://www.gcc-sg.org/en-us/CognitiveSources/DigitalLibrary/Lists/DigitalLibrary/Water%20and%20Electricity/1414489603.pdf

Global Water Intelligence (2020) 32nd Worldwide Desalting Plant Inventory Media Analytics Ltd.

Government of Canada (2020). Impact Assessment Act. (S.C. 2019, c. 28, s. 1). Current to 2020-07-28 and last amended on 2019-08-28. https://laws.justice.gc.ca/eng/acts/I-2.75/page-1.html#h-1160082.

Government of Mauritius. (2005). EIA guidelines for proposed desalination plants. Department of Environment. Ministry of Environment and National Development Unit. http://environment.govmu.org/English/eia/Documents/eia/desguide.doc.

Grobelak A., Czerwińska K. and Murtaś A. (2019) General considerations on sludge disposal, industrial and municipal sludge. In: Industrial and Municipal Sludge, M. N. V. Prasad, P. J. de Campos Favas, M. Vithanage and S. V. Mohan (eds.), Elsevier, Oxford, UK, pp. 135–153.

Guang L., Koomson D. A., Jingyu H., Ewusi-Mensah D. and Miwornunyuie N. 2020. Performance of Exoelectrogenic Bacteria Used in Microbial Desalination Cell Technology. *International Journal of Environmental Research and Public Health*, **17**, 1121. doi:10.3390/ijerph17031121.

Hancock N. T., Black N. D. and Cath T. Y. (2012). A comparative life cycle assessment of hybrid osmotic dilution desalination and established seawater desalination and wastewater reclamation processes. *Water Research*, **46**, 1145–1154. DOI: 10.1016/j.watres.2011.12.004

Harford A. J., Trenfield M. A., Cheng K. L. and van Dam R. A. (2014), Ecotoxicological assessment of manganese. Internal report 630. Departmetn of the Environment, Australia. https://www.environment.gov.au/system/files/resources/7eb03536-e1b6-4ad3-8469-b07a7ece7295/files/ir630.pdf.

Hoepner T. (1999). A procedure for environmental impact assessments (EIA) for seawater desalination plants. *Desalination*, **124**, 1–12.

Hristozov D., Gottardo S., Semenzin E., Oomen A., Bos P., Peijnenburg W., van Tongeren M., Nowack B., Hunt N., Brunelli A., Fordsmand J. J. S., Tran L. and Marcomini A. (2016). Frameworks and tools for risk assessment of manufactured nanomaterials. *Environment International*, **95**, 36–53.

Ibrahim H. D., Elfatih A. B. and Eltahir E. A.B. (2019). Impact of Brine Discharge from Seawater Desalination Plants on Persian/Arabian Gulf Salinity. *Journal of Environmental Enginerring*, **45**(12), 04019084. DOI: 10.1061/(ASCE)EE.1943-7870.0001604.

International Association for Impact Assessment [IAIA]. (1999). Principles of environmental impact assessment best practice. International Association For Impact Assessment, Institute of Environmental Assessment, UK. https://www.iaia.org/publications.

International Association for Impact Assessment [IAIA]. (2009). What Is Impact Assessment? https:\\www.iaia.org\publications.

International Atomic Energy Agency [IEAE]. (2010). Environmental impact assessment of nuclear desalination. Report IAEA-TECDOC-1642. Vienna, Austria.

International Energy Agency [IEA]. (2019). Desalinated water affects the energy equation in the Middle East. IEA, Paris. https://www.iea.org/commentaries/desalinated-water-affects-the-energy-equation-in-the-middle-east.

International Institute for Sustainable Development. (2020). EIA: 7 Steps. https://www.iisd.org/learning/eia/eia-7-steps/.

International Standards Organization [ISO]. (2006). ISO 14044:2006(en). Environmental management — Life cycle assessment — Requirements and guidelines. Geneva.

IPCC. (2019). IPCC Special Report on the Ocean and Cryosphere in a Changing Climate. In: H.-O. Pörtner, D.C. Roberts, V. Masson-Delmotte, P. Zhai, M. Tignor, E. Poloczanska, K. Mintenbeck, A. Alegría, M. Nicolai, A. Okem, J. Petzold, B. Rama, N.M. Weyer (eds.). © 2019 Intergovernmental Panel on Climate Change. Geneva, Switzerland. https://www.ipcc.ch/srocc/download/.

Irlbeck M. J. and Voutchkov N. (2013). Desal Dialog: A Regulatory Workshop on Critical Issues of Desalination Permitting. Water Reuse Research Foundation, Water Research Foundation, Bureau of Reclamation, US Department of the Interior.

Jamtsho S. (2019). Life Cycle Assessment of Microbial Desalination Cell for Low Energy Drinking Water from Seawater. Master's Degree Thesis, IHE Delft Institute for Water Education, Delft. Netherlands.

Jones E., Qadir M., van Viet M. T. H., Smakhtin V. and Kand S. (2019). The state of desalination and brine production: A global outlook. *Science of the Total Environment*, **657**, 1343–1356.

Kabir E., Kumar V., Kim K.-H., Yip A. C. K. and Shon J. R. (2018). Environmental impacts of nanomaterials. *Journal of Environmental Mangement*, **225**, 261–271.

Keller A. A., Garner K., Miller R. J. and Lenihan H. S. (2012). Toxicity of Nano-Zero Valent Iron to Freshwater and Marine Organisms. *PLoS One*, **7**, e43983.

Kenigsberg C., Abramovich S. and Hyams-Kaphzan O. (2020). The effect of long-term brine discharge from desalination plants on benthic foraminifera. *PLoS One*, **15**(1), e0227589. https:// doi.org/10.1371/journal.pone.0227589.

Law N., Ansari S., Livens F. R., Renshaw J. C. and Lloyd J. R. (2008). Formation of nanoscale elemental silver particles via enzymatic reduction by *Geobacter sulfurreducens*. *Applied and Environmental Microbiology*, **74**(22), 7090–7093. doi:10.1128/AEM.01069-08.

Lawler W., Alvarez-Gaitan J., Leslie G. and Le-Clech P. (2015). Comparative life cycle assessment of end-of-life options for reverse osmosis membranes. *Desalination*, **357**, 45–54.

Lenntech B. C. (2020). Reverse Osmosis Desalination: Brine disposal. https://www.lenntech.com/processes/desalination/brine/general/brine-disposal.htm#:~:text=Brine%20concentration%20varies%20from%2050,and%20any%20related%20human%20activities.

Lerner K. L. (2020). Desalination, The Gale Encyclopedia of Science. https://www.encyclopedia.com/science-and-technology/technology/technology-terms-and-concepts/desalination-water

Manju S. and Sagar N. (2017). Renewable energy integrated desalination: A sustainable solution to overcome future fresh-water scarcity in India. *Renewable and Sustainable Energy Reviews*, **73**, 594–609.

Mehanna M., Basséguy R., Délia M. L. and Bergel A. (2009). Effect of *Geobacter sulfurreducens* on the microbial corrosion of mild steel, ferritic and austenitic stainless steels. *Corrosion Science*, **51**, 2596–2604.

Mekonnen M. M. and Hoekstra A. J. (2016). Four billion people facing severe water scarcity. *Science Advances*. 2016, **2**(2), e1500323.

Melbourne Water GHD. (2007). Melbourne Augmentation Program. Sea water desalination. Feasibility Study. Melbourne, Australia. ISBN 0977585867.

Michaud L. D. (2017). Ferrocyanide toxicity. 911 Metallurgist. https://www.911metallurgist.com/blog/ferrocyanide-toxicity.

Monterey Bay National Marine Sanctuary (MBNMS). (2010). Guidelines for Desalination Plants in the Monterey Bay National Marine Sanctuary. National Marine Fisheries Service and National Oceanic and Atmospheric Administration, Monterrey CA. USA.

Muñoz I. and Fernández-Alba A. R. (2008). Reducing the environmental impacts of reverse osmosis desalination by using brackish groundwater resources. *Water Research*, **42**, 801–811. DOI https://doi.org/10.1016/j.watres.2007.08.021

National Snow and Ice Data Center [NS&IDC]. (2020). Salinity and Brine. https://nsidc.org/cryosphere/seaice/characteristics/brine_salinity.html last updated Apr 3 2020.

Nebbia G. and Menozzi G. M. (1966). A short history of water desalination. Federazione delle Associazioni Scidentifiche e Techniche. Axqua Dolce dal mare. 11th Inchiesta Internazionale. Milano, April 18–19, pp. 129–175. https://www.researchgate.net/publication/301328310_A_short_history_of_water_desalination

Opher T. and Friedler E. (2016). Comparative LCA of decentralized wastewater treatment alternatives for non-potable urban reuse. *Journal of Environmental Management*, **182**, 464–476. DOI http://dx.doi.org/10.1016/j.jenvman.2016.07.080

Ozair G., Al-Sebaie K. Z. and Al-Zahrany S. A. (2017). Impact of long term concentrated brine disposal on the ecosystems of nearshore marine environment – a case study. The International Desalination Association World Congress on Desalination and Water Reuse 2017/Sao Paulo, Brazil. REF: IDAWC17_58002.

Park J. G., Lee B., Jo S. Y., Lee J-S. and Jun H-B. (2018). Control of accumulated volatile fatty acids by recycling nitrified effluent. *Journal of Environmental Health Science and Engineering*, **16**, 19–25. https://doi-org.unesco-ihe.idm.oclc.org/10.1007/s40201-018-0291-9

Peters A., Crane M., Maycock D., Merrington G., Simpson P., Sorokin N. and Atkinson C. (2010). Proposed EQS for Water Framework Directive Annex VIII substances: manganese (total dissolved). Bristol, United Kingdom: Environment Agency.

Petersen K. L., Paytan A., Rahav E., Levy O., Silverman J., Barzel O., Potts D. and Bar-Zeev E. (2018). Impact of brine and antiscalants on reef-building corals in the Gulf of Aqaba e Potential effects from desalination plants. *Water Research*, **144**, 183–191.

PRODES. (2010). Guidelines for the regulation of desalination. Promotion of Renewable Energies for Water Production through Desalination (PRODES) Project. https://www.prodes-project.org.

Ramírez-Moreno M., Rodenas P., Aliaguilla M., Bosch-Jimenez P., Borràs E., Zamora P., Monsalvo V., Rogalla F., Ortiz J. M. and Esteve-Núñez A. (2019). Comparative Performance of Microbial Desalination Cells Using Air Diffusion and Liquid Cathode Reactions: Study of the Salt Removal and Desalination Efficiency. In: International Society for Microbial Electrochemistry And Technology: Outputs From The 2018 Regional Meetings. Frontiers in Energy Research. S. Glaven, U. Schröder,E. H. Yu, S. Mutnuri, J. A. Gralnick and S. H. Chan (eds.). Frontiers Research Topics. Lausanne, Switzerland, pp. 216–227. doi.org/10.3389/fenrg.2019.00135.

Ran J., Wu L., He Y., Yang Z., Wang Y., Jiang C., Ge L., Bakangura E. and Xu T. (2017) Ion exchange membranes: New developments and applications. *Journal of Membrane Science*, **522**, 267–291.

Saccà M. L., Fajardo C., Costa G., Lobo C., Nande M. and Martin M. (2014). Integrating classical and molecular approaches to evaluate the impact of nanosized zero-valent iron (nZVI) on soil organisms. *Chemosphere*, **104**, 184–189.

Sadeghi L., Yousefi Babadi V. and Espanani H. R. (2015). Toxic effects of the Fe2O3 nanoparticles on the liver and lung tissue. *Bratislavske lekarske listy*, **116**, 373–378.

Santa Cruz Biotechnology. (2010). Potassium hexacyanoferrate (III) sc-203355, Safety Data Sheet. https://datasheets.scbt.com/sc-203355.pdf

Shaw B. J. and Hardy R. D. (2011). Physiological effects of nanoparticles on fish: A comparison of nanometals versus metal ions. *Environment International*, **37**, 1083–1097

Shipman H. R. (1974). Water Desalination. Central Projects Staff. Public Utilities Department. Report PUN-009. February 20 International World Bank for Reconstruction and Development.

Shrestha P. (2020). A guideline for Environment and Social Impact Assessment of Reverse Osmosis desalination and Microbial Desalination Cells. MSc. thesis. IHE Delft Institute for Water Education. Delft, Netherlands.

Singh S. P., Kumari M., Kumari S. I., Rahman M. F., Mahoob M. and Grover P. (2013). Toxicity assessment of manganese oxide micro and nanoparticles in Wistar rats after 28 days of repeated oral exposure. *Journal of Applied Toxicology*. 2013, **33**, 1165–1179.

Swartz C. D., du Plessis C. A., Burger A. J. and Offringa G. (2006). A desalination guide for South African municipal engineers. *Water SA*, **32**(5), 641–647. http://www.wrc.org.za

Tarnacki K., Melin T., Jansen A. and Medevoort J. (2011) Comparison of environmental impact and energy efficiency of desalination processes by LCA. *Water Science Technology: Water Supply*, **11**, 246–251.

United Nations Environment Programme [UNEP]. (2007). Desalination Resource and Guidance Manual for Environmental Impact Assessments. United Nations Environment Programme, Regional Office for West Asia, Manama, and World Health Organization, Regional Office for the Eastern Mediterranean, Cairo.

United Nations Environment Programme [UNEP]. (2017). Decision IG.23/13. Updated guidelines on the management of desalination activities. UNEP (DEPI)/MED IG.23/23. https://wedocs.unep.org/handle/20.500.11822/22569.

United Nations Environment Programme [UNEP]. (2019). Environmental Rule of Law. First Global Report. United Nations Environment Programme, Nairobi. https://www.unenvironment.org/resources/assessment/environmental-rule-law-first-global-report.

Wang J., Liu C., Ji J., Luo R., Hu X., Sun X., Shen J., Han W. W. and Wang L. (2017). In-situ incorporation of iron-copper bimetallic particles in electrospun carbon nanofibers as an efficient Fenton catalyst. *Applied Catalysis B: Environmental*, **207**, 316–325.

Warheit D. B. (2018). Hazard and risk assessment strategies for nanoparticle exposures: How far have we come in the past 10 years? *F1000Research*, **7**, 376. DOI: 10.12688/f1000research.12691.1.

Water Research Foundation. (2010). Guidelines for Implementing Seawater and Brackish Water Desalination Facilities. Water Research Foundation and Arsenic Water Technology Partnership. ISBN 978–1-60573-130-8.

Wiesner M. R., Lowry G. V., Jones K. L., Hochella M. F., Jr., Di Giulio R. T., Casman E. and Bernhardt E. S. (2009). Decreasing Uncertainties in Assessing Environmental Exposure, Risk, and Ecological Implications of Nanomaterials. *Environmental Science & Technology*, **43**, 6458–6462.

World Atomic Association. (2020). Desalination. Updated March 2020. https://www.world-nuclear.org/information-library/non-power-nuclear-applications/industry/nuclear-desalination.aspx

World Business Council for Sustainable Development [WBCSD] and the Cement Sustainability Initiative [CSI]. (2016). Guidelines for environmental & social impact assessment.16) https://www.wbcsd.org/Sector-Projects/Cement-Sustainability-Initiative/Resour ces/Guidelines-for-Environmental-and-Social-Impact-Assessment-ESIA.

World Health Organization [WHO]. (2007) Desalination for safe water supply: guidance for the health and environmental aspects applicable to desalination World Health Organization (WHO), Geneva, Switzerland.

Zaitseva N. V., Zemlyanovaa M. A., Zvezdina V. N., Akafievab T. I. and Saenkoc E. V. 2015. Acute Inhalation Toxicity of Manganese Oxide Nanoparticles. *Nanotechnologies in Russia (Rossiiskie Nanotekhnologi)*, **10**(5–6), 468–474.

Zhang Y.-Q., Dringen R., Petters C., Rastedt W., Köser J., Filserd J. and Stolte S. (2016). Toxicity of dimercaptosuccinate-coated and un-functionalized magnetic iron oxide nanoparticles towards aquatic organisms. *Environmental science Nano*, **3**, 754–767.

Zhang J., Yuan H., Deng Y., Zha Y., Abu-Reesh I. M., He Z. and Yuan C. (2018). Life cycle assessment of a microbial desalination cell for sustainable wastewater treatment and saline water desalination. *Journal of Cleaner Production*, **200**, 900–910. DOI https://doi.org/10.1016/j.jclepro.2018.07.197.

Chapter 8

New frontiers and outlook

Juan M. Ortiz[1], *Sergio G. Salinas-Rodríguez*[2], *Nathan Bossa*[3], *Eduard Borràs*[3], *Juan Arévalo*[4] and *Abraham Esteve-Núñez*[1,5]

[1]*IMDEA Water Institute, Avenida Punto Com, 2, Parque Científico Tecnológico de la Universidad de Alcalá, 28805, Alcalá de Henares, Madrid, Spain*

[2]*IHE Delft Institute for Water Education, Water Supply, Sanitation and Environmental Technology Department, Westvest 7, 2611 AX, Delft, Netherlands*

[3]*LEITAT Technological Center, C/Pallars, 179–185, 08005 Barcelona, Spain*

[4]*FCC Aqualia S.A., Avenida del Camino de Santiago, 40, Building 3, 4th floor, 28050, Madrid, Spain*

[5]*Department of Analytical Chemistry, Physical Chemistry and Chemical Engineering Department, Universidad de Alcalá, Alcalá de Henares, Spain*

ABSTRACT

The main goal of this chapter is to outline the areas for further development and research of microbial desalination cells (MDCs) for production of low energy drinking water. This chapter highlights the challenges ahead in the context of membrane and electrode development, scaling-up, construction, environmental challenges, capital and operational costs, social acceptancy and further identification of niche areas for MDCs.

Keywords: Challenges, costs, further research and development, outlook, social acceptancy, state of the art

doi: 10.2166/9781789062120_0221

8.1 INTRODUCTION

As indicated in chapter 1 of this book, the simultaneous negative issues related to water stress and energy depletion have led in recent decades to the research of exploring effective solutions for water desalination with less energy consumption. Microbial desalination cell (MDC) technology is a recent technology that could contribute to the sustainability of desalination as well as to decreasing the energy costs of conventional reverse osmosis (RO) plants. From the author's point of view, the research and innovation carried out in the MIDES project represents an ambitious effort to bring the technology into real application and gain insights to overcome the main technological barriers for market implementation. Nevertheless, there are some issues that should be addressed before the real practical application of MDCs, and then, future efforts should be focused in order to demonstrate the economic and technical feasibility of MDC technology, as indicated in the next sections.

8.1.1 Membrane development

As extensively described in chapter 3, ion exchange membranes are key elements of MDC technology, and in general of all microbial electrochemical technologies. As the currently available membranes in the market have been designed and produced for other areas of application (i.e., energy systems, electrodialysis for separation processes), it would be desirable to continue developing specific ion exchange membranes for desalination in bio-electrochemical devices. In this sense, apart from improving current biofouling resistance performance, special attention should be paid to improving properties such as low back diffusion rates, water loss and production costs, as well as mechanical stability.

8.1.2 Electrode development

Electrodes, including current collectors, are also key elements of MDC systems, and significantly contribute to the total capital costs of the system. For this reason, special carbon materials for bio-electrochemical devices must lower current production costs while maintaining performance for real deployment of the technology. Efforts in research of innovative materials with improved microbial adhesion to the surface or a custom porous structure for microbial colonization could bring new advances for microbial desalination with respect to anodes.

8.1.3 Air diffusion cathode scale-up and construction

As described in chapters 2 and 3, MDCs operating with an air diffusion cathode are suitable for desalination of brackish water and seawater and the MIDES project has demonstrated the use of low cost materials for an oxygen reduction reaction (ORR) with sufficient performance to provide electric potential to drive the desalination process (i.e., iron doped carbon nanofibres). This concept was scaled-up and tested in a pre-pilot MDC system. Future actions could be focused on the development of this concept at large scale for pilot-plant validation and performance in long-term duration experiments.

8.1.4 Catholyte regeneration (MDC with liquid catholyte strategy)

When liquid catholyte is used in MDC technology, an increase in the desalination performance is obtained, as described in chapter 2. However, liquid catholyte has to be regenerated when depleted, and this represents the main drawback of this strategy. For this reason, it could be convenient in further studies to establish an easy and cost-effective procedure for regeneration of depleted catholyte solution, including the integration of renewable energy (i.e., photovoltaic, wind energy) or coupling microbial electrochemical reactions (i.e., biocathodes) for regeneration.

8.1.5 Organic substrate availability

The advantages of wastewater treatment in MDC systems must also be considered, as the MDCs are a combined technology of both water desalination and wastewater treatment. It is obvious that MDC plants should be constructed in a location near wastewater and saline water sources. In this sense, the application of MDC technology for reclaimed water for irrigation represents a rational application of the technology in wastewater plants, when a desalination process is required due to high salinity of the effluent. However, the limitation of organic substrates could be a negative factor for the implementation of MDC systems, especially in isolated or remote areas. In this sense, the study of alternative organic substrates for MDC operation could be an interesting subject to expand the concept of sustainable desalination using MDC (for example: microalgae, forest residues, etc.).

8.1.6 Capital and operation costs

The MDC reactor could have a complex assembly that may require more consideration of design, construction, maintenance and operation. In addition, the MIDES concept includes pre- and post-treatments such as ultrafiltration of saline water, anaerobic digestion of wastewater, reverse osmosis, remineralization and disinfection. Currently, MDC technology development has reached pilot-plant scale, and validation could bring new information for subsequent design and optimization of commercial prototypes. For this reason, further efforts should also be focused on construction optimization, reduction of costs in materials and coupling systems, looking for production of economically viable solutions based on MDC technology.

8.1.7 Nanomaterials safety consideration

As described in chapter 7, the risk associated with the use of nanomaterials has been validated at the lab-scale. The risk was considered minimal and no specific safety action was required. Safety has been assessed during the air-cathode use and end-of-life, but the potential risks to the worker during the air-cathode synthesis were not determined. So, this aspect should be considered in the future. Additionally, in case of future commercialization, since 1 January 2020, companies must provide more information on nanomaterials in the EU market under the REACH Regulation (ECHA/NR/19/35), so this aspect must be taken into account to fulfil the regulation requirements.

8.1.8 Social acceptance of the technology

MDC systems take advantage of organic substrates in water solution to obtain energy to drive the desalination process. Thus, the simultaneous production of fresh water and wastewater treatment is achieved. In most cases, MDC uses organic substrate which is contained in wastewaters (i.e., urban or industrial). High quality, safe drinking water is essential for public health and well-being and an important asset for the economy. In practice, safe water could be obtained from the MIDES concept, as RO and disinfection are implemented as post-treatment, ensuring the quality of treated water, even if minor leakage or cross-contamination occurs in the microbial desalination cell. However, the social perception of production of drinking water by using wastes could be an issue for commercial development of MDC technology. For this reason, measures to increase social acceptance of the technology could be important for the future of the technology.

8.1.9 New niches for MDC technology application

Potential use of MDC goes far beyond the desalination purposes tackled in MIDES. MDC can be an alternative technology to treat industrial saline waters from electroplating, mining and petrochemical industries, opening the door to its use for electro-bioremediation purposes. In addition, MDC may be used for selectively recovering high-added value resources from wastewaters such as nutrients, volatile fatty acids and metals. These alternative MDC applications might increase the circularity of key resources from industrials wastewater streams.

8.2 SUMMARY

MDC technology has promising potential as a sustainable process for water desalination, compared to more conventional technologies of desalination, that is, RO or thermal processes. MDCs could be either employed as a stand-alone technology for water treatment and recycling or could also be integrated with the traditional membrane-based RO plants as a strategy for reducing energy requirements by decreasing salinity of the feed water. The MIDES project has developed and explored MDC technology for brackish water and seawater desalination using different approaches and achieving clear developments in MDC components such as membranes, electrodes, hardware, control and automatization systems, mathematical modelling, etc. Finally, future works focusing on the currently described technology constraints will contribute to bringing forward the implementation of microbial desalination and new ideas for a more sustainable future in the water sector.

Abbreviations and acronyms to Chapter 7

BSE	Back-scattered-electron
CEM	Cation exchange membrane
CEMC	CEM Corporation
CNF	Carbon nanofibers
COD	Chemical oxygen demand
CWB	California Water Board
EC50	Half maximal effective concentration
EDX	Energy Dispersive X-ray spectroscopy
EIA	Environmental impact assessment
ENMs	Engineered nanomaterials
ESIA	Environmental and social impact assessment
g/L	Grams per litre
GCC	Gulf Cooperation Council
GDL	Cathode gas diffusion layer
GSGCC	General Secretariat Gulf Cooperation Council
GDE	Gas Diffusion Electrode
G. Mu	Ministry of Environment, Mauritius
GEPAP	Gulf Environmental Partnership and Action Program
Hoep	Hoepner
HRSEM	High-resolution Scanning electron microscopy
IAEA	International Atomic Energy Agency

doi: 10.2166/9781789062120_0227

IAIA	International Association for Impact Assessment
IEA	International Energy Agency
IISD	International Institute for Sustainable Development
ICP-MS	Inductive plasma mass spectrometry/atomic emission spectroscopy
IONP	Iron oxide nanoparticles
kDa	Kilo Daltons
kDa	Kilo Dalton
kWh/m^3	Kilowatt hour per cubic meter
LCA	Life cycle assessment
LCI	Life cycle inventory
m^3/d	Cubic meter per day
MARS	Microwave [acid] reaction system
MBNMS	Monterey Bay National Marine Sanctuary
MDC	Microbial desalination cell
MIDES	Microbial Desalination for Low Energy Drinking Water
MSF	multi-stage flash
MDC	Microbial desalination cell
NEP	Nano-enabled product
NHMRC	National Health and Medical Research Council
NRMMC	Natural Resource Management Ministerial Council
NMs	Nanomaterials
nM	nanometre
NS&IDC	National Snow and Ice Data Center
nZVI	Nanoscale zero-valent iron
p.p.m.	Parts per million
PBS	Phosphate-buffered saline
PNEC	Predicted non effect concentration
PRODES	Promotion of Renewable Energies for Water Production through Desalination
RGM	Resource and Guidance Manual
RO	Reverse osmosis
SEM	Scanning electron microscopy
SGL	SGL Group
S-ZA	Swartz *et al.*
µM	micrometre
UF	Ultrafiltration
UNEP	United Nations Environment Programme
W	Watt
WBCSD	World Business Council for Sustainable Development
WHO	World Health Organization
WRF	Water Research Foundation
WWT	Wastewater treatment

Annex to Chapter 7

Annex 1 Potential impacts resulting from the operation of the MDC[1]

MDC Component	Environmental Component	Impact
Wastewater (Intake/Storage)	Groundwater quality	Contamination of groundwater by discharge or spills of wastewater.
	Seawater quality	Contamination of seawater by discharge of waste-water or sludge
Wastewater (Intake/Storage)	Human health	Unpleasant odours from wastewater leaks, causing discomfort of workers and/or people living around the plant
		Perceived or actual risk of contamination of desalinated water by wastewater
Wastewater (effluent)	Quality of surface water	Contamination of surface by the discharge of wastewater effluents
		Risk of eutrophication
	Marine organisms	Mortality of aquatic species and disruption of biotic interactions

(*Continued*)

doi: 10.2166/9781789062120_0229

Annex 1 Potential impacts resulting from the operation of the MDC[1] (*Continued*).

MDC Component	Environmental Component	Impact
Iron air-cathode (nano particles)	Workers health	Risk of inflammation of lung tissues
		Increased risk of numbers of neutrophils, eosinophils, and lymphocytes in lung tissue
		Risk of inflammation of liver tissues
	Marine organisms	Risk of mortality of marine organisms
Manganese oxide air-cathode (nano particles)	Workers health	Risk of neurological effects in workers
		Risk of psychological disorder
		Risk of lungs and tumour necrosis
Ferro-ferricyanide	Human health and marine organisms	Lethal effects on humans and aquatic organisms if it reacts in the environment and produces hydrogen cyanide
	Marine organisms	Mortality of algae and protozoa
		Risk of mortality of marine organisms
Electrode biofilm	Piping	Corrosion of steel
Cleaner for ceramic membrane (sodium hypochlorite)	Workers health	Risk of skin irritation and burns
		Risk of eye problems in occupational workers
Cleaner for ceramic membrane (hydrogen peroxide)		Risk of bronchitis or pulmonary oedema
		Risk of bleaching or skin whitening
		Risk of eye problems
	Marine organisms	Alteration of the community structure of bacterio-plankton

(*Continued*)

Annex 1 Potential impacts resulting from the operation of the MDC[1] (*Continued*).

MDC Component	Environmental Component	Impact
Cleaner for ceramic membrane (hydrochloric acid)	Workers health	Risk of irritation to the respiratory tract, skin, nose, eyes and pulmonary oedema
	Marine organisms	Mortality of marine organisms due to increase of pH
Cleaner for ceramic membrane (sodium hydroxide)	Workers health	Risk of severe burns, blistering, permanent scars and dermatitis
		Risk of severe damage to the cornea of the eyes
	Marine organisms	Mortality of marine organisms due to increase of pH
Ion exchange membrane	Solid waste	Pollution of landfills due to periodic replacement of membranes
Anode/cathode		Pollution of landfills due to periodic replacement of membranes
Ceramic membranes		Pollution of landfills due to periodic replacement of membranes

[1]From Grobelak *et al.* (2019), Mehanna *et al.* (2009) and Shrestha (2020).

IWA Publishing's authorised EU representative for General Product Safety Regulations is Diane D'Arras, 15 rue Duret, 75116 Paris, France, e-mail: safety@iwap.co.uk.

Printed and bound by CPI Group (UK) Ltd, Croydon, CR0 4YY
16/06/2026
02140372-0001